George Kordis

In the Joy of Cinnabar

Sebastian Press, 2025

GEROGE KORDIS

IN THE JOY OF CINNABAR

Published by
St. Sebastian Orthodox Press

Contemporary Christian Thought, number 104

Edited by
Bishop Maxim Vasiljević

Design & Prepress
Denis Vikić

Printing
Birograf COMP
Belgrade, Serbia

Address all correspondence to:
St. Sebastian Orthodox Press
1621 West Garvey Avenue
Alhambra, California 91803

Email: info@westsrbdio.org
Website: www.westsrbdio.org

Names:	Vasiljević, Maksim, 1968- editor.
Title:	George Kordis : in the joy of cinnabar / edited by Bishop Maxim Vasiljević.
Description:	Alhambra, California : St. Sebastian Orthodox Press, [2025] | Series: Contemporary Christian thought series ; no. 104. | Includes bibliographical references.
Identifiers:	ISBN: 9781964233130
Subjects:	LCSH: Kordēs, Giōrgos, 1956- | Icons, Byzantine--Technique. | Icon painting--Technique. | Color in art. | Icon painting--Byzantine influences--Technique. | Tempera painting--Technique. | Painting, Byzantine--Technique.
Classification:	LCC: N8189.5.K6 V37 2025 | DDC: 759.9495--dc23

George Kordis

In the Joy of Cinnabar

Edited by
Bishop Maxim Vasiljević

St. Sebastian Orthodox Press

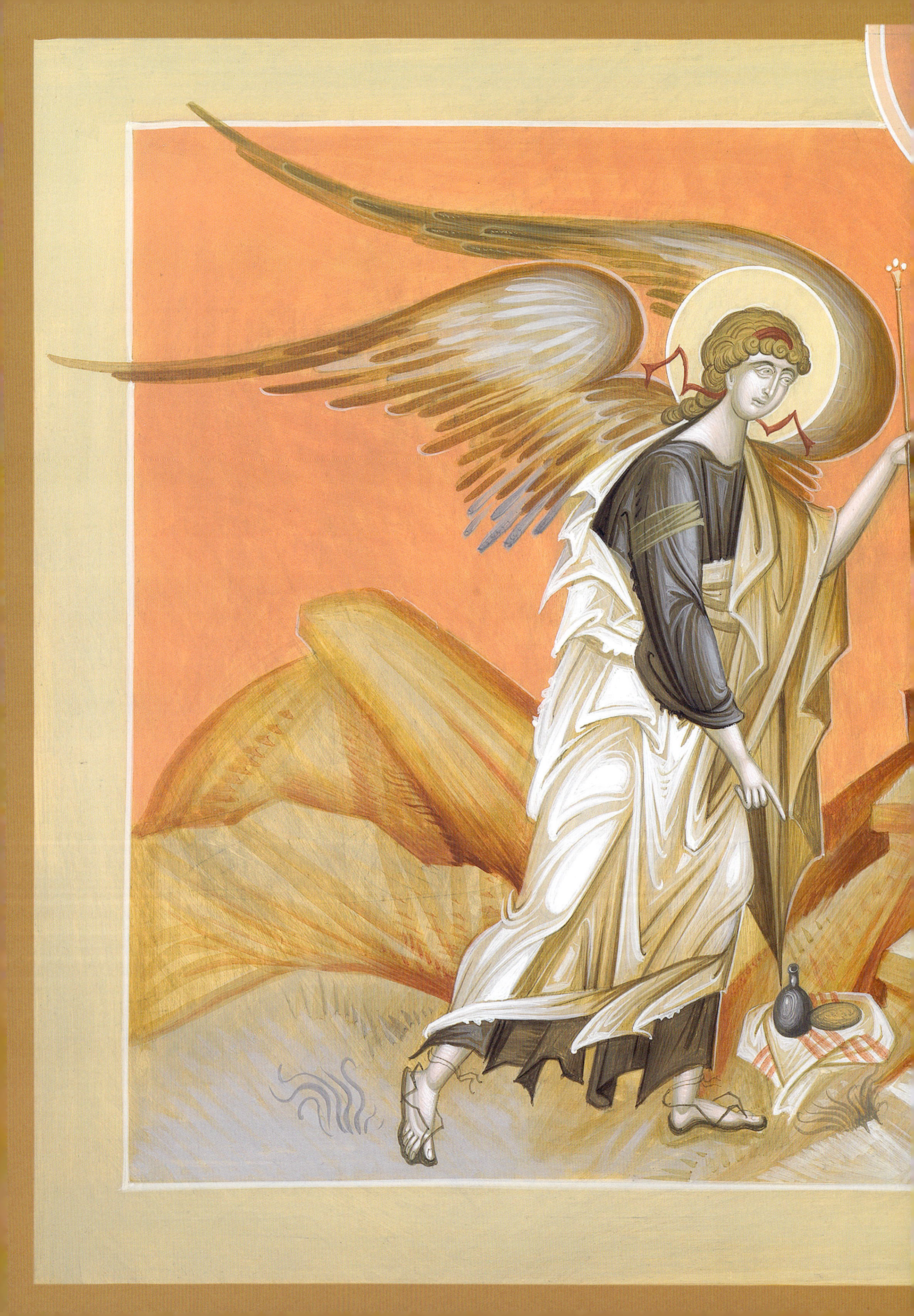

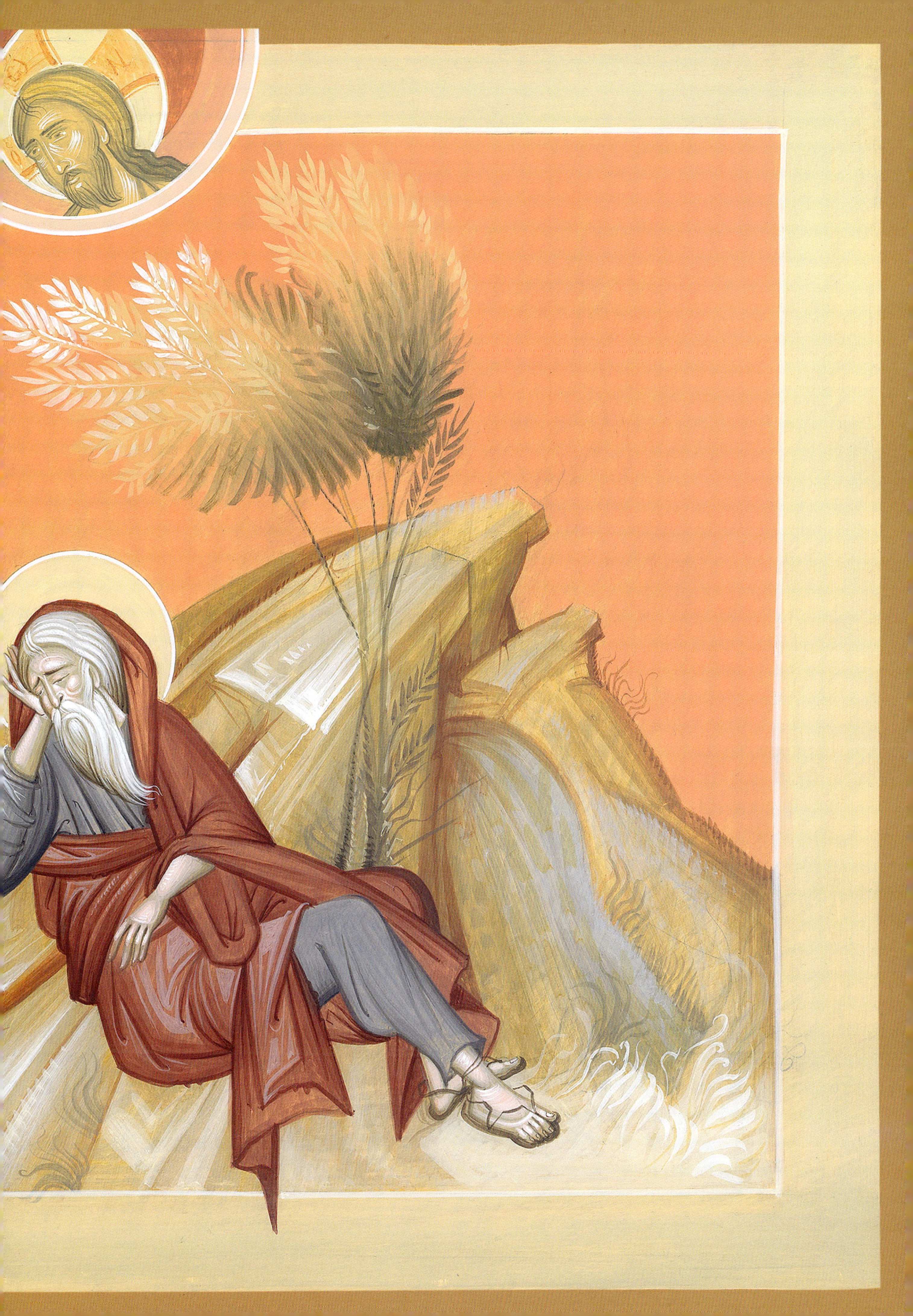

ANGEL FEEDING PROPHET ELIAS IN DESERT,
egg tempera on wood, 35 x 55 cm

Contents

Foreword: Byzantine Presence, Modern Breath

Bishop Maxim Vasiljević

The art of George Kordis offers a compelling invitation to explore the dynamic interplay between tradition and innovation, the sacred and the contemporary. Kordis, acclaimed globally for his iconography, is not merely a practitioner of an ancient art form; he is a scholar, a theologian, and a visionary who reimagines Byzantine artistic principles for the modern world.

I first encountered the work of George Kordis in the early 1990s, when his iconography—resplendent, rhythmic, and deeply rooted—began to appear in Orthodox circles as both a renewal and a revelation. From the first moment, I was drawn to the dynamic energy of his compositions. His lines did not merely frame holy figures—they pulsed. They danced. They beckoned. His icons carried the weight of centuries and the breath of now.

Over the decades, I have watched his oeuvre grow into a body of work that is both unmistakably Byzantine and unmistakably Kordis. The dynamic quality of his compositions—neither static nor mimetic—conveys the living tradition of Orthodox art as something profoundly theological, deeply personal, and inherently relational.

This volume provides a unique opportunity to engage with Kordis' work on multiple levels. As you turn these pages, you will encounter icons that resonate with the visual language of Byzantium: the abstract forms, the vibrant colors, the emphasis on line, and the absence of spatial recession. Yet, you will also discover an artist deeply conversant with modernism, unafraid to infuse his work with a powerful personal style that embraces contemporary aesthetics.

Kordis' genius lies in his ability to bridge seemingly disparate worlds. He draws inspiration from the "relational ontology" of contemporary Orthodox theologians like Christos Yannaras and John Zizioulas, emphasizing relationship as the motive force in Byzantine art. This theological grounding informs his understanding of the icon as "an image woven between the wall and the viewer," an invitation to communion and encounter.

The texts presented here illuminate various facets of Kordis' artistic vision. We see his mastery of color, used not merely decoratively but as a "means of expression and creating structure," where warm hues draw the viewer in and cool shades create depth. We learn how he employs "contrapposto" and rhythm to create dynamic compositions, drawing inspiration from ancient Greek art and infusing his figures with a sense of movement and internal life. We also see how Kordis personifies nature, imbuing landscapes with an almost anthropomorphic quality, further enhancing the expressive power of his work. Beyond the formal elements, these texts explore the rich tapestry of allusions in Kordis' art: to religion, philosophy, Greek and international poetry, literature, and folklore. His work becomes a "visual poetry" that balances the past and the present, the local and the universal, the sacred and the everyday.

Ultimately, the art of George Kordis is an invitation to see anew. It is a call to engage with tradition not as something static, but as a living, breathing entity capable of constant renewal. It is an art that seeks to communicate, to transform, and to offer a glimpse into the eschatological. As you explore the work of George Kordis within these pages, prepare to be challenged, inspired, and drawn into a world where the beauty of the past illuminates the possibilities of the future.

Art historians and theologians have illuminated this vitality in rich and varied ways. Annemarie Weyl Carr eloquently situates Kordis within the lineage of Byzantine art while highlighting the theological depth of his practice, particularly his engagement with the "relational ontology." In her care, Kordis becomes not only a master iconographer but also a visual theologian who gives shape and color to communion itself. Peter C. Bouteneff praises Kordis for understanding that "tradition is creativity," a phrase that distills the paradox of faithfulness and innovation so essential to the iconographic task. Judith Wolfe, with poetic precision, places Kordis alongside George Herbert, capturing how his art turns space into presence, rhythm into prayer. Justin Willson presents George Kordis as an iconographer who harmonizes Byzantine tradition with modern visual language, using color, form, and rhythmic composition to create spiritually charged narratives that radiate hope, beauty, and theological depth. And Niobe Kritikou gives us a comprehensive and contemplative portrait of the artist, tracing the philosophical, historical, and technical layers that give his work both gravitas and grace.

To behold a Kordis icon is not simply to observe a scene—it is to participate in a rhythm of divine-human communion, to enter into a mystery that is always unfolding. And in this, perhaps, lies his greatest gift to our time: the recovery of iconography not as a preserved style of the past but as a living act of theology, drawn with reverence and freedom.

This volume honors that gift. May it draw each reader closer not only to the art of George Kordis, but to the living iconography of Christ, whose beauty ever calls us into deeper communion.

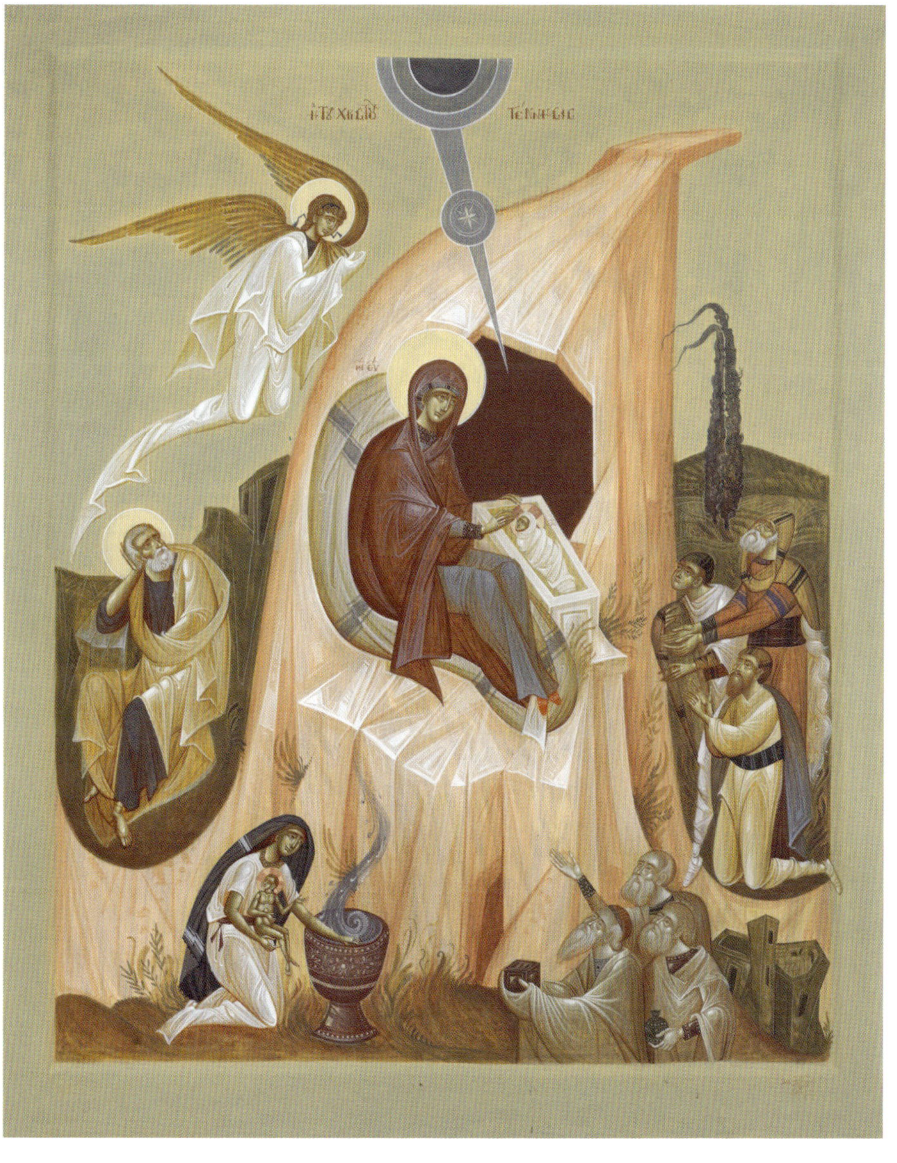

CHRIST'S NATIVITY, egg tempera on wood, 80 x 100 cm

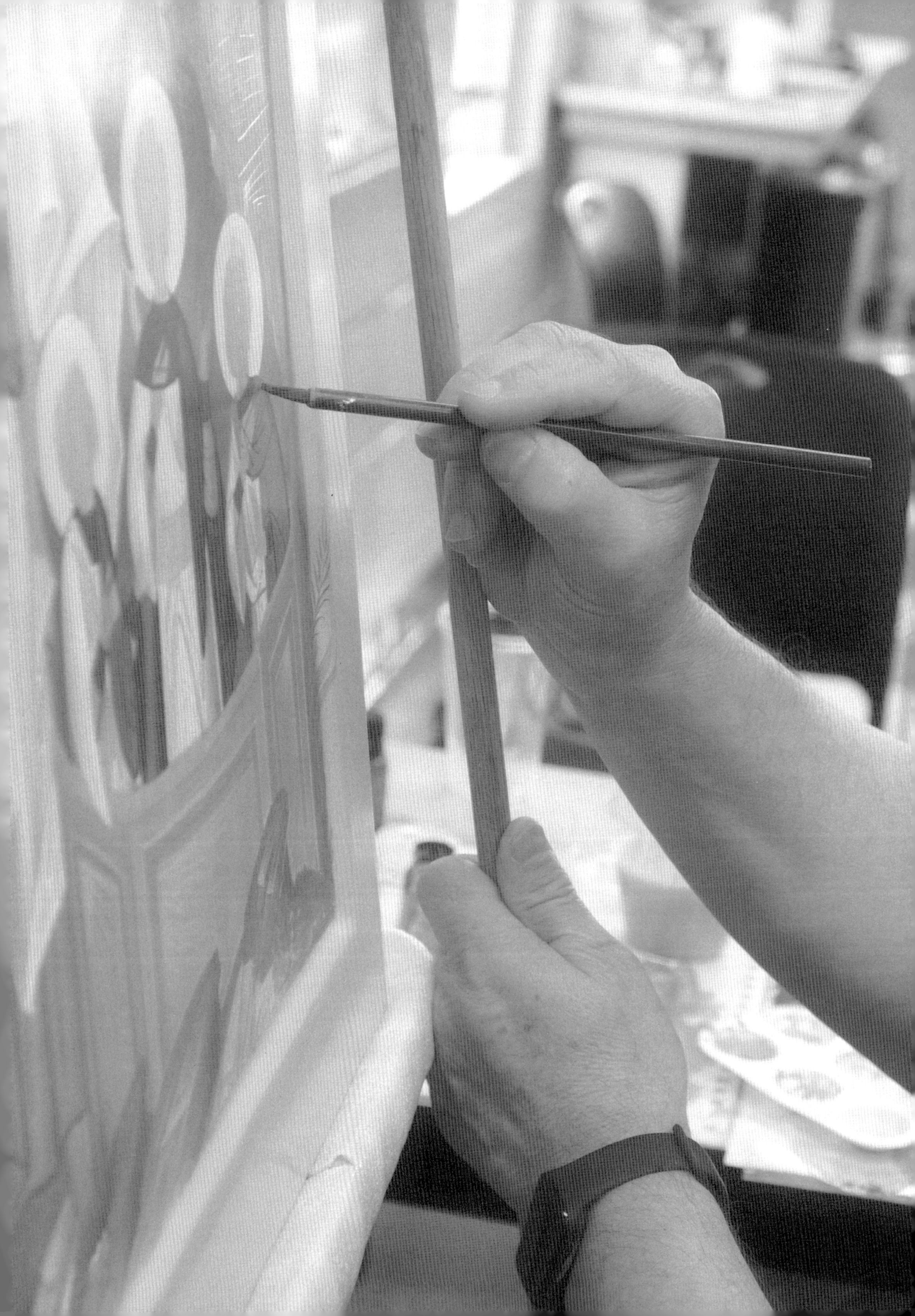

The Byzantine Art of George Kordis

Annemarie Weyl Carr

This exhibition of icons by the acclaimed Greek iconographer George Kordis offers visitors to the Icon Museum an arresting opportunity to observe the work of an esteemed, present-day iconographer in direct juxtaposition with the traditional works that the Museum's patrons know well. Kordis' works are recognized throughout the globe: he has frescoed congregational churches, exhibited his portable icons, and shown his secular paintings across the modern world. But he is also a scholar deeply conversant with the theology, image theory, and artistic practice of Orthodoxy since its origins in early Byzantium.[1] He has written compelling books on the practice of icon painting, delineating its fundamentals and demonstrating how it serves to mediate a philosophy of sacred order that is rooted in the writings of the early Byzantine Fathers. The art of the icon, he writes, is Byzantine art.

Two factors make his books especially rewarding. One is the language in which he analyses the formal effects distinctive to Byzantine icons, using the terms developed in the wake of modernism to describe the relation of forms to both the picture plane and the viewing eye. This makes their visual effects accessible in the terms not of subjective response but of recognized aesthetic analysis. The second gift of his books is the way they align the formal effects of Byzantine art with the "relational ontology" developed by Christos Yannaras and John Zizioulas in contemporary Orthodox theology.[2] These thinkers have sought to reassert within Orthodox practice its definitive spirit of ecclesial communion, of union without dissolution of individual identity expressed especially formatively in writings on the Trinity the fourth-century Greek Fathers. The paradox of a communion in which each member is both

[1] See especially, in English, Icon as Communion: *The Ideals and Compositional Principles of Icon Painting*, trans. Caroline Makropoulos (Brookline, MA: Holy Cross Orthodox Press, 2011), and *Color as Light in Byzantine Painting: Theory and Practice with a Guide to Egg Tempera with Underpainting*, trans. Karen D. Symons (Alhambra, CA: St. Sebastian Orthodox Press, 2020).

[2] John D. Zizioulas, *Being as Communion: Studies in Personhood and the Church*, Contemporary Greek Theologians Series 4 (London: T.T. Clark, 1997); Idem, *Communion and Otherness: Further Studies in Personhood and the Church*, w. Paul McPartlan (London: T.T. Clark, 2007); Christos Yannaras, *Relational Ontology*, trans. Norman Russell (Brookline, MA: Holy Cross Orthodox Press, 2011); Idem, *Person and Eros*, trans Norman Russell (Brookline, MA: Holy Cross Orthodox Press, 2008).

individually unique and lovingly participant in the whole serves as a model both for Divinity itself, in the Trinity, and for relationships within the church, which unites diverse members within the one body of Christ. To be in the church is to be in relationship with Christ, with his saints, with one another, and with the world. Relationship—of one element to another within an image, of image to viewer, and of viewer to subject seen in the image—is the motive force shaping Byzantine art as Kordis analyzes it. "What I therefore intended in all my art," he has written, "was to record a system of relationships. How the saints related to Christ, how they related to creation, to other people, and to us, the viewers of their icons, who are called to be saints. ... All those relationships had to become shapes and colors so that they could become an aesthetic experience for the viewers who communed with the icons."[3]

His icons certainly display the characteristics most readily associated with icons. They are abstract, brightly colored, linear, and without spatial recession. But these essentially stylistic features are not what make them icons. Kordis does not regard icon painting as a style. No less than easel painting, icon painting has nurtured countless different styles. Kordis' own icons have a powerfully individual style inflected with modern abstract sensibility.[4] They make no attempt to adopt or recreate the style of any one iconographer or period from the Byzantine past. What makes his works icons is their fidelity to an underlying system of representation with its own rules of recognizability, its own conventions for expressing space, light and plasticity, its own rhythmic demands, and its own ways of relating viewers visually to its images. It is a dynamic system, that both binds its forms to one another and moves them forward from the picture plane toward the viewer. It is not just the forms on the surface but the relation viewers assume in response to them that gives the icon its singular impact. The icon, Kordis says, "is an image woven between the wall and the viewer,"[5] and its representational conventions abet this. Thus, Byzantine artists avoid suggesting depth behind the picture plane, since depth draws the figures away from the viewer. Rather than receding, bodies fracture into segments along the surface, to be recomposed as rhythmically integrated wholes. Their volumes are modeled vigorously in three dimensions, but not in light and shadow, or in accord with a coherent source of cast light as if caught in a particular moment. Thus liberated from the transience of momentary effects, the figural forms assume less appearance than presence, in all its concerted force, and unfold before us in front of the picture plane. Their closeness invites response, drawing the viewer into relation with the subject. As Evan Freeman writes, "For Kordis, Byzantine-style icons are not, above all, non-naturalist and therefore spiritual, but rather, they are relational in order to be ecclesial. Their purpose is to unite all people as the body of Christ."[6]

The icons Kordis has gathered in this exhibition include no static, frontal half-lengths, and no golden backgrounds. Instead, they show fig-

[3] George Kordis, "The Light of the World. Pictorial Solace with Icons of Saints of Mount Athos," in *Το Φως του Κόσμου. Μια ζωγραφική Παραμυθία με ιστορημένους Αγίους του Αγίου Όρους, The Light of the World. Pictorial Solace with Icons of Saints of Mount Athos*, exhibition catalogue accompanying "Light of Lights. Athos. Byzantium. Ecumene," Thessaloniki, Mount Athos Center, 22 June – 15 September 2017 (Thessaloniki: Mount Athos Center, 2017, 52-55 at 54.

[4] As Thanassis Semoglou writes in "Icons of Prayer," in Kordis, *Το Φως του Κόσμου*, 38-39, at 39: "The meeting of Orthodox religious art with early twentieth-century European artistic trends in Kordis' work . . . dynamically reinstates the concept of abstraction in Byzantine art."

[5] Kordis, *Icon as Communion*, 51.

[6] Evan Freeman, "The Relational Icons of George Kordis and Byzantium," in *Light from Light. Icons 2002–2022*, ed. Lambros Kamperides (Montreal: George Kordis, and Alexander Press, 2022), 37-42, at 40.

ures in scenes of active relationship, their gestures conveying often fervent interconnectivity across the luminous surface of the panel. Perhaps the most classic of these fervent confrontations is that between the hieromonk Zosimas and the spectral Mary of Egypt, as she receives the Eucharist from his hand.[7] The Eucharistic communion of Mary of Egypt, the flamboyant harlot smitten by the Virgin Mary into decades of ascetic remorse, recurs often in icon painting, though rarely with the haunting beauty of this version. Its temperature is set already by the burning tenderness of the background color, against which the blue of the Jordan River throbs darkly. The two colors blend in the twilight purple of Zosimas' vestment, while the rose flames into pure red in the Eucharistic morsel that he extends. The morsel's convergence with the brim of Mary's halo stops time as truly as the breathless stillness of a frontal saint does. But here the still point sits amid surging lines of respondent energies. The taut line of the confronted gazes, the beams that spread to embrace both faces, the rightward ascent of Zosimas' arm, the ardent leftward answer of Mary's bending back, and beneath it the urgent extension of her pleading hands work together to create a grid of interactive energies. When Zosimas first saw Mary, he thought he was seeing a specter; swathed here in a tracery of white, she seems indeed pure energy, a vortex of light and line. Line, for Kordis, "distinguishes [Byzantine art] from any other tradition. ... In Byzantine art, all form is movement, everything is energy ... [T]his energy—which in the end becomes rhythm, is the fundamental characteristic of Byzantine art."[8] Byzantine art is traditionally called static; to Kordis it is dynamic: a forcefield of inter-responsive energies.

The paired hands that Mary of Egypt had extended to Zosimas to receive the Eucharistic morsel recur in the icon of the Prophet Elijah, who extends his hands to receive the morsel of bread or meat delivered by the raven. Like Zosimas and Mary of Egypt, Elijah and the Raven (1 Kings 17: 3-4) was a theme familiar in Byzantine iconography, where sustenance wondrously provided in the wilderness was as much an affirmation of ascetic renunciation as it was a celebration of the prophet himself. The ascetic life is evoked here by the cave with its dangling basket. Rarely, however, was Elijah shown so precipitously elevated as here, as if conflating his nourishment by a raven with his later ascetic withdrawal at Horeb on Mount Sinai (1 Kings, 19: 8-18). This emphasizes his role as a great prophet and seer, an identity reinforced by his fabled fur mantle. His gaze, like Mary of Egypt's, is fixed far beyond the morsel, evoking the deity whom Elijah himself had identified as a "small voice," both intimate and beyond visualization (1 Kings 19: 11-12). The prophet's entire posture is absorbed in his gaze. Both the diagonal of the mountain peak and the isolation of his head against the background suggest that he is alone in a vast space. His powerful limbs swell into the ample distance, taking commanding place within it. They are fascinatingly modeled in a paradoxical play of pure color, as their advancing portions are

[7] The Life of Mary of Egypt was written down in the seventh century by Saint Sophronius, Patriarch of Jerusalem, some hundred years after her death: stmaryofegypt.org/files/library/life.htm (consulted 19 January 2025).

[8] Kordis, *Icon as Communion*, 7.

darker in tone than their "shadowed" undersides, which glow brightly. Rather than adhering to the canons of Renaissance modelling, in which shadowed areas darken and lose color, they adhere to the logic of Byzantine art, where form never loses color. In this, the receding surfaces of a volumetric form are warm in tone while the advancing ones are cool. In Elijah, the cool tones declare the garments' local color, which is deep and strong, as if in full light, while the receding surfaces glow with ocher, as if bathed in ambient light. At no point does form fall into darkness; advancing or receding, it radiates with light and color. Nor, however, does it ever really recede into depth. If Elijah's head seems deep in space, his right foot is if anything even closer to us than the frame just below it. In fact, if we follow the rhythmic contours into which his swelling limbs are packed, they flow smoothly upward against the picture plane, forcing the volumes forward. As Kordis, the consummate colorist, tells us, "with the appropriate perspective drawing, the order of colors and their tonal transition, the icon is cast into the viewer's space-time, enters their space-time and becomes the present."[9]

The icons of Mary of Egypt and Elijah show how critical line and color are to Kordis' art; so is rhythm. A revealing example of the power of rhythm is seen in his treatment of the scene of the Supper at Emmaus, in which the newly resurrected Christ revealed himself to two unsuspecting disciples named Cleopas and Luke (Luke 24: 13-35). This scene is usually composed symmetrically, and indeed, between the parentheses of the two disciples, the eye settles here on the composition's central chord, ascending from the blue chalice to the morsel clasped against Christ's familiar blue mantle, and from there to the similarly blue cusp of the roof above him. But then unexpectedly, the cusp swerves off-axis to the left, triggering a rhythmic readjustment of all the forms around it. The disciple at left rises slightly from his seat to capture and complete the descending arc that the roof had begun; the inward curve of his bent back is echoed in the arc of the tree above him and continued in the canted hilltop at right; the flanking buildings skew their roofs to emphasize the disciples' up-turned faces. This ripple of response turns attention to the area within it. It is occupied by the head of Christ, and he, too, unexpectedly, turns toward the left. His gaze is met by the paired hands and ardent face of the rising disciple. With much the same gesture, Mary of Egypt had gazed toward Zosimas as he brought her the Eucharistic morsel; Elijah, too, had gazed with extended hands and upraised face toward his unseen and unimaginable God. The disciple here gazes on God, as well, but now as a responsive, human person. The very buildings they occupy, the landscape around it, and the figures' lightly destabilized symmetry all participate to rivet attention on the interchange between Christ's and his disciple's face. Among the key themes of Christos Yannaras' responsive ontology is, in Rowan Williams' words, "to establish that the relation between God and man is personal and reciprocal, a relation of communion, a 'real' (as opposed to a logical)

[9] Kordis, "The Light of the World," 55.

relation."[10] Kordis' rhythmic rethinking of Christ's self-revelation at Emmaus offers a riveting image of the experience of God as a profoundly personal one, with a responsive fellow person. The embracing rhythm extends even to the ripples of the tablecloth exposed by the left disciple's rise, as they open a path for the viewer into the being-in-response of the icon's central figures.

The Supper at Emmaus is one of the eleven Resurrectional Gospels read at Matins after Easter, but it is not one of the twelve major feasts and so is not among the themes most often taken up in panel painted icons. Far rarer is the scene of Saint Andrew recognizing Jesus as the Messiah, recounted in John 1: 41-42. Andrew heard John the Baptist extoll Jesus and so told his brother, Peter, that "we have found the Messiah, which is, being interpreted, Christ." Together, they went as disciples to Jesus; it was at this meeting that Jesus called Peter "Cephas," a rock. Andrew's recognition of Jesus is only very rarely pictured in any medium. In Byzantine art, it appears in the "frieze Gospels," rare Gospel manuscripts that illustrate almost every event in the text; in Western medieval art, it is included in the similarly dense cycle illustrating the *Meditationes Vitae Christi*, but otherwise it is not recorded. Thus Kordis can have had little precedent to guide him, and he creates a composition presenting Andrew and Peter quite literally against the backdrop of their lives as fishermen in the Sea of Galilee. The sea rises behind them, its blue surface in some sense hallowing them as if in a mandorla and differentiating them as a pair from the taller, paler Christ, whose form rises against the open ground plane. He alone in the scene is haloed and identified by name. Above him, in place of a title, are the words of the Gospel: "Andrew first found his brother and said we have found the Messiah." As in Kordis' other icons, the figures are dynamic and interrelated. Andrew is in the middle, reaching with one hand to his brother while gesturing animatedly to Christ. Where Peter's posture is uncomprehending, Andrew's is intense, fixed on Christ. Though the biblical text above him speaks only about "finding" the Messiah, Andrew's pose is clear: he has *seen* that this is Christ. This gives the event a special resonance. Theologians, especially Byzantine theologians and artists preoccupied by issues of images' validity, have given keen attention to biblical passages that speak of seeing (or not seeing) God—the first chapter of John's Gospel, which includes this episode, is especially rich in them.[11] Similarly, scenes of seeing gained special resonance in art. Kordis' icon makes Andrew's recognition one of these. No less than Symeon at the Presentation in the Temple—or Luke and Cleopas in Emmaus—Andrew here verifies the visibility of God as man. That visibility, in turn, validates the art of the icon.

Among the notable features of the icon of Andrew's recognition is its enclosure within a frame of pure black. This is exceptional. Like easel paintings, icons traditionally do have frames, since their images occupy recessions, framed by the panel into which they are carved. As such, the

[10] See Rowan Williams, "The Theology of Personhood: A Study of the Thought of Christos Yannaras," *Sobornost* 6 (1972): 415-30, repr. in *Polis, Ontology, Ecclesial Event: Engaging with Christos Yannaras' Thought*, ed. Sotiris Mitralexis, Andreas Andreopoulos, Pui Him Ip, Isidoros Katsos, and Dionysios Skliris (London: The Lutterworth Press, James Clarke & Co Ltd., 2018), 226-242, at 235.

[11] See John 1: 14: "And the Word was made flesh, and dwelt among us, and we beheld his glory. . .;" John 1: 18 "No man hath seen God at any time; the only begotten Son, which is in the bosom of the Father, he hath declared him;" John 1: 34: "And I saw [at the Baptism], and bare record that this is the Son of God;" John 1: 36: "And looking upon Jesus as he walked, he saith, Behold the Lamb of God!" It was this last statement that convinced Andrew to accept Jesus as the Messiah, which is to say, Christ.

frames are integral to the icon, not independent impositions upon it. They are characteristically painted with the same color as the image's background, and suggest inclusion rather than separation. Art historians have been intrigued by the way Byzantine painters used frames, and Kordis, too, has clearly pondered them. Not only has he experimented with the black frame here, but he has painted icons without frames, like the two scenes here of the Embrace of Anna and Joachim, the parents of the Virgin Mary, and the embrace of the pregnant Mary with her elderly cousin, Elizabeth, also pregnant. Known as the Visitation, the latter is recounted in Luke 1: 39-44; the first is apocryphal rather than biblical but is a customary component of Mary's life in Byzantine art. A miracle accompanied each embrace: Mary was conceived in the first, and the embryonic John the Baptist leapt in Elizabeth's womb at the approach of the embryonic Christ in the Visitation. Thus each is a hallowed event. Nonetheless, they stand apart from the other icons. This is partly because the closure of the figures' postures precludes using their limbs to lace the composition together in a grid of gestural energy. Instead, the setting becomes exceptionally assertive, exposing Kordis' skill as an abstract painter. But no less significant is the absence of a frame in each. These deeply human interchanges, which occur between two people, lack the preparatory interval of the frame, in which the viewers can gather their thoughts to enter holy ground.

Frames reappear in the three further icons, including the narrow border around the softly-cadenced scene of human and animal effort with the Holy Family's Flight into Egypt. Aside from the crimson of Mary's tiny slippers, the color reserved in Byzantium for royalty, the scene includes only the very slightest hint that the effortful journey is divinely guided. This emerges in the way the angel's wing just overlaps the rising bevel of the frame. This shows that the scene exceeds its appearance: the heaven implied in the Child's upward gaze flows over the frame and into the picture. Larger, more forceful frames appear in the final two icons. The scenes within them are explosive. The sacrifice of Isaac, known in Byzantine art as the Sacrifice of Abraham (Genesis 22: 1-14), is among the Bible's most gripping dramas, and Kordis embraces its dynamism, composing it with two energetically coiled, convergent forms. Clenched around his son's body, Abraham twists upward to raise his arm for the blow; it is seized by the angel, who corkscrews downward across the implicit boundary between heaven and earth, image and frame, to stop him. But it is hard to say exactly what is within and what is outside the scene here. As Kordis has emphasized, Byzantine art is designed to move the figures not back into receding depth but forward toward the viewer, and Abraham's plastically modeled forms do press forward from the softly colored ground; moreover, they spill out over the frame, treating it more as a form of background than as a superimposed limit. The frame itself, however, refuses to settle back as background. It is more like a foreground: it cuts off our view of the mountains; its color is high and bright and heavenly; and the angel must catapult into Abraham's sacrifice from his own, native realm of heaven. Yet frame and background remain distinct. The frame's function is spelled out by Isaac's sandals, set in it as if carefully put off before entering holy ground. The frame is a threshold area, setting off the special context within which the holy takes place.

A similarly ample, if slightly dimmer, frame sets off the figure of Saint Maximus the Confessor as he struggles to compose his greatest book, known as the *Mystagogy,* or "induction into mystery." Maximus was a seventh-century monk and theologian, canonized not as a martyr, though he was brutally tortured for his theological stances, but as a great mind. He is among the Fathers most revered by theologians today. He is seen in icons, but almost always as a frontal half-length figure. This passionate image of him immersed in his work is exceptional. The blue sea behind him may reflect his wide travel, since he is known to have spent periods of his life in Constantinople, Palestine, North Africa, and Rome; on the other hand, it may function together with the scrolls at his feet as a reference to the two companion volumes often linked with the *Mystagogy,* of which one examined the natural world and the other the Bible.[12] The mystery into which the *Mystagogy* initiates us is that of the Divine Liturgy: the book is a dense interpretative exposition of the Christian service of divine worship. It opens with a description of the physical church building in which the liturgy is performed and then moves on to unfold the spiritual depth of the ritual performance, first for the clergy and their lay congregants, and thereafter for monks and contemplatives, whose life of withdrawal may preclude regular communion, as had been the case with Mary of Egypt. It is monastics, including a woman very like Mary, whom Kordis depicts engaged in the worship that Maximus describes. They extend their hands toward the angels, with whose heavenly worship the human liturgy merges mystically when performed. The God to whom they turn in worship is Christ, who responds deeply and personally, as his turned head shows. Maximus closes his text with a summary, presenting its three-part structure as a foundation for two fundamental points. One is the mutual conformity of the perceptible forms of physical creation with their metaphysical ground in God, the one informing the other, as the physical form of the church building exposes the spiritual form of the ritual it serves. The second point is the union without dissolution of individual identity that is forged by the church, as Christ binds his infinitely varied individual devotees as one body to himself, who is their head. This way of being-in-relation was modeled for the church by the union in diversity of the Trinity, a parallel that would later be developed in the relational ontology of Christos Yannaras, who affirmed that "Maximus the Confessor fascinated me—the result, *Person and Eros*."[13]

Saint Maximus the Confessor was the close friend and disciple of Saint Sophronios, Patriarch of Jerusalem (634-638) and author of the

12 Especially clearly laid out by Fr. Maximos Constas in "Contemplating Christ in the Church: the *Mystagogy* of St Maximos the Confessor" at https://www.pappaspatristicinstitute.org/post/mystagogy.intro (viewed 19 January 2025).

13 Chrisros Yannaras, *Καθ' εαυτόν* (Athens: Ikaros, 2005), 50, quoted by Norman Russell in his introduction to Christos Yannaras and Norman Russell, *Metaphysics as a Personal Adventure: Christos Yannaras in Conversation with Norman Russell*, Orthodox Christian Profiles Series 9 (Yonkers, NY: St. Vladimir's Seminary Press, 2017).

haunting story of Zosimas and Mary of Egypt. Kordis' evocation of Mary in his icon of Saint Maximus is in this sense absolutely apt. Moreover, his icon of Zosimas and Mary illustrates in action a major message of Maximus' *Mystagogy*. With its double explication of the liturgy, one directed to the clergy and their lay congregants, and one directed to ascetic contemplatives, Maximus conveyed his conviction that the Divine Liturgy should be the path to Christ of both groups: Christians in the world and Christians in monastic withdrawal. It was the monastics whom Kordis chose to depict engaged in liturgical worship in his icon of Saint Maximus; in turn, in his icon of Mary of Egypt, the fiercely reclusive Mary receives at last the salvific gift of the Eucharist from the hands of a monk who is also a priest. By different means, the two icons make a convergent point: that the Eucharist is the path to union with Christ for both monks and lay people, however individually divergent they are. Icons are often called theology in visual form. That is just what the icon of Saint Maximus is, a visualization of a theological point. But icons are also visual invitations into experiences of God. It is such scenes of being-in-relation that dominate Kordis' current exhibition.

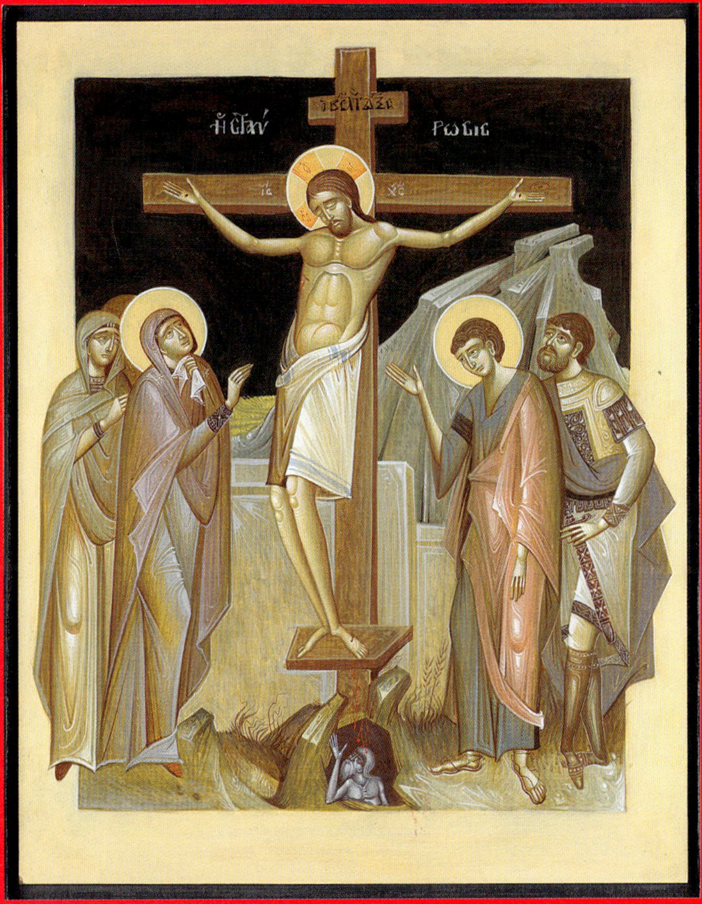

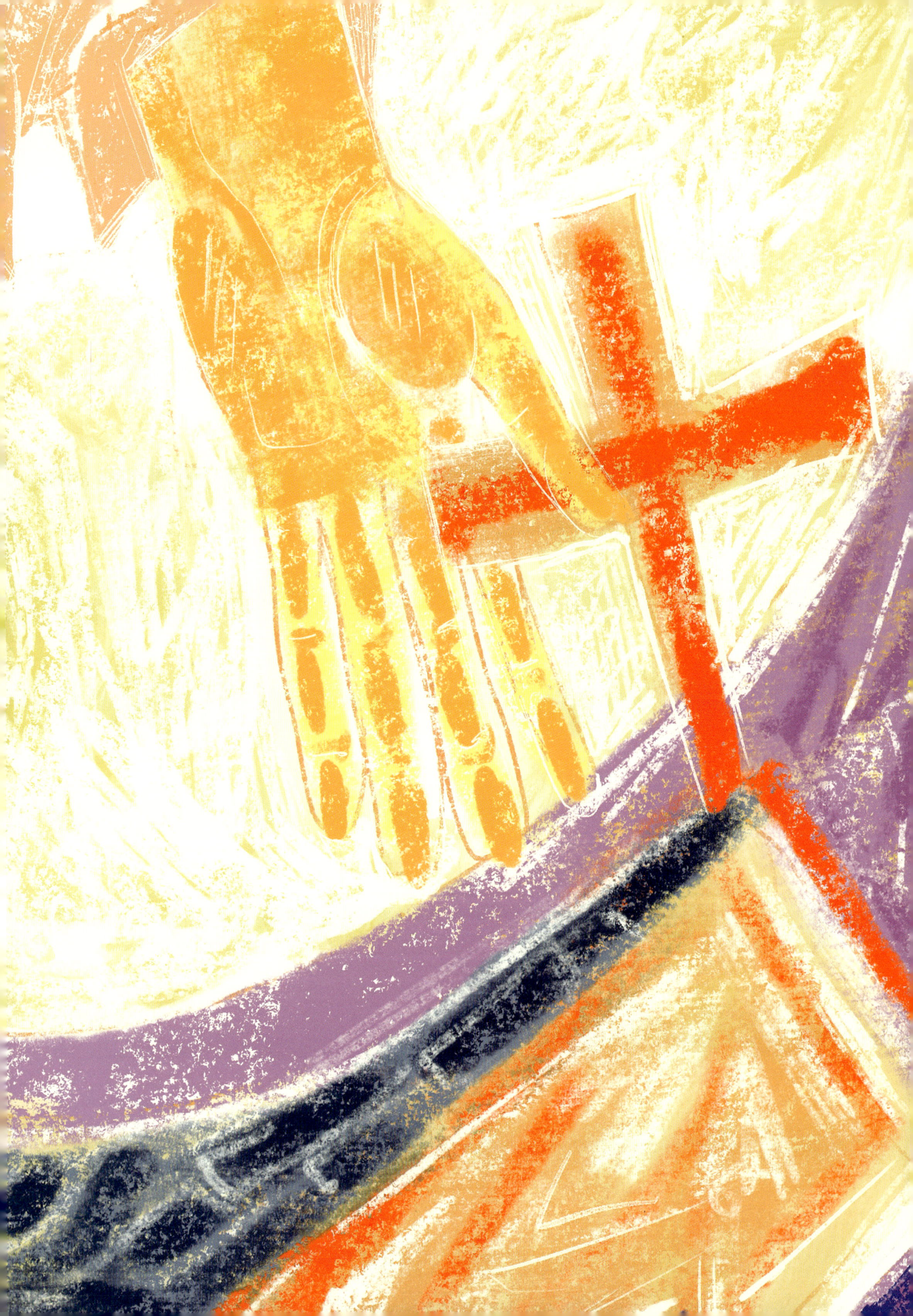

George Kordis: A Contemporary Iconographer

Peter C. Bouteneff

George Kordis is one of the most important living iconographers in the Orthodox world. But let us give some substance to this simple fact.

His significance derives from the character and level of his contribution to iconography in several dimensions. The most evident is the art itself. Kordis' iconography spans a range of media: frescoes, portable icons, and installations. In any of these, the beholder is liable to find works that hew closely to traditional forms and subjects and depictions that alternately delight and challenge expectation (or both at once). Several trademarks stem from his artistic vision: his use of color, his depiction of light, his sense of motion and line, his understanding of visual dynamics, all deployed with exceptional technical skill.

And then there are lovely—and sometimes challenging—surprises: The lines on the floor defy any pattern or symmetry! Look, there's a little bird in the fountain, a ship on the sea in the background, a fellow in ... blue jeans? Such details give us but one hint that the icon we are beholding came from Kordis' brush.

But what we encounter in a Kordis icon is finally the artist's understanding of the nature of "tradition." For some ecclesiastical artists and audiences, "tradition" exists in a tension (if not a war) with "creativity." Some circles, in fact, frown on the word "creativity," since it seems to denote self-indulgence and a flight from rightful obedience to "tradition."

On the basis of his own intuition as well as his careful, extensive, and thoughtful study of ecclesiastical and cultural history, Kordis understands this relationship differently: "Tradition *is* creativity." The happy fact is that however some may struggle with such a formulation, when we encounter George's actual work—in an Athonite monastery, on the walls of a church in Philadelphia, in an icon of a saint or a feast—it sweeps us up into a spirit of reverence, humility, and joy.

This, at any rate, is my experience of his work. If anyone still finds obstacles on the way to that place of devotion and wonder, it is not because Kordis aims to test, to push, or to persuade anyone. No. He is merely God's faithful instrument in the Church's creative work. His integrity, his gifts, and his vision are such that this work is reverential, theologically "true," and exquisitely beautiful.

Real Presences

Judith Wolfe

Exalted manna, gladness of the best,
Heaven in ordinary, man well drest,
The milky way, the bird of Paradise,
Church-bells beyond the stars heard, the soul's blood,
The land of spices; something understood.

George Herbert, "Prayer (I)"

George Kordis says that what we call tradition is really creativity: a perpetual movement, carrying the past into the present and on towards the future. It is a dance, patterned by hereditary steps but animated by free, expressive energy.

Icons make present the saints that they depict. Kordis suggests that this presence is not brought about merely by the force of the pictorial forms (which are, in fact, in danger of being empty shells) or the direct inspiration of the Holy Spirit. It is also brought about by the co-creative agency of the artist: an agency that is at once active and passive, a gathering and a letting go. The artist studies, observes, practices, and then gets out of the way, letting the lines flow through him and following their demands. This simultaneous self-investment and self-emptying allows the saints he paints to become present, to live with the artist's own life as well as their own. It also invites those who behold the icons to gather and empty themselves in turn, to encounter the saint in the depth of their own life. The life of this creative energy radiates from the icons of this exhibition and suffuses the space around them, making that space for the viewers, as prayer does for George Herbert, a "land of spices," a "heaven in the ordinary."

Three of the icons in this exhibition form a visual group of sorts: Abraham, Elijah, and Maximus the Confessor. All three tend towards the monochrome, and feature a central, aged male figure looking upwards with a raised arm to the viewer's right. Abraham looks towards his raised, knife-bearing hand, lightly arrested by an angel; Elijah looks towards a bird descending from the sun and raises his arms to receive its food; and Maximus fixes his gaze on a vision of Christ adored by angels and saints as he raises his hand to write his *Mystagogy*.

Each of these icons exemplifies tradition as movement, not by any self-conscious "updating" but by entering into the inner life of the scenes and manifesting this life both as it happens in time and as it looks in the light of eternity.

Kordis' Abraham is profoundly moving. He kneels with his son Isaac pressed against him. Isaac pityingly, almost lovingly looks down at the ram. Abraham holds Isaac with one hand and raises the knife with the other in a reverse grip that is ready to stab upwards, as if he is just as ready to pierce himself as his son, or himself along with his son. The angel is a mirror image of Abraham's contorted posture. Both exhibit a form of *contrapposto*, lower and upper half turned in opposite directions. But where Abraham's is agonized, the angel's is light and floating. Where Abraham holds a knife, the angel holds a sceptre. In Kordis' iconography, angels are often *rhythmic* forms, their postures expressing a divine rhythm that echoes through creation. This angel sets a counterpoint to Abraham's movement and interrupts his desperate obedience with the rhythm of Annunciations that stretches from Genesis 3 to Luke 1.

Kordis' Elijah is perhaps the most luminous of all the icons exhibited here. The scene burns in a bright yellow, Elijah's hair flowing in the fiery draft of the sun, his halo flaming around him, his whole being illuminated. The movement of the raven and the glow of the bread (with recalls the live coal of Isaiah 6) cascades on through Elijah's mantle of hair. All this brightness is counterposed by a hole in the rock, revealing a dark cave or well with a bucket suspended. This may recollect the widow of Zarephath, whom Elijah asked for water, and whose provisions never ran out—a story itself later consummated in Jesus' promise to the Woman at the Well that "the water that I will give will become in [those who drink it] a spring of water gushing up to eternal life." Here, the icon gives a sense of Elijah drawing from a deep, dark well of waiting and suffering, through which he has learnt how to receive the gift of the heavenly food now offered to him.

Kordis' Maximus is closer to us viewers than the patriarch and the prophet: like us, he encounters the divine in a vision rather than by corporeal touch. The classic iconographic pose and inscription of Christ's head in the upper right-hand corner identifies iconography with mystical sight—it is structurally unclear whether Maximus is gazing at an icon of Christ or at a mystical vision. Whether corporeal or incorporeal, the vision is shared by an enveloping throng of saints and angels. They are behind Maximus, out of his sight, but they envelop him in a movement of shared attention in which everything in the icon participates: an upward movement towards Christ of all eyes, all arms, and even salamander, fish, quill and scrolls. A rhythm of movement pulses through creation and unites it in shared adoration. This is a vision of the church—the church as building, as liturgy, and as community, filled with mystical life.

These icons make visible some of the invisible rhythms of creation—interruption, acceptance, adoration—and draw us to experience and participate in them. In doing so, they also carry on the rhythm of tradition, a living movement which Kordis gathers, allows to flow through him, and passes on to the future.

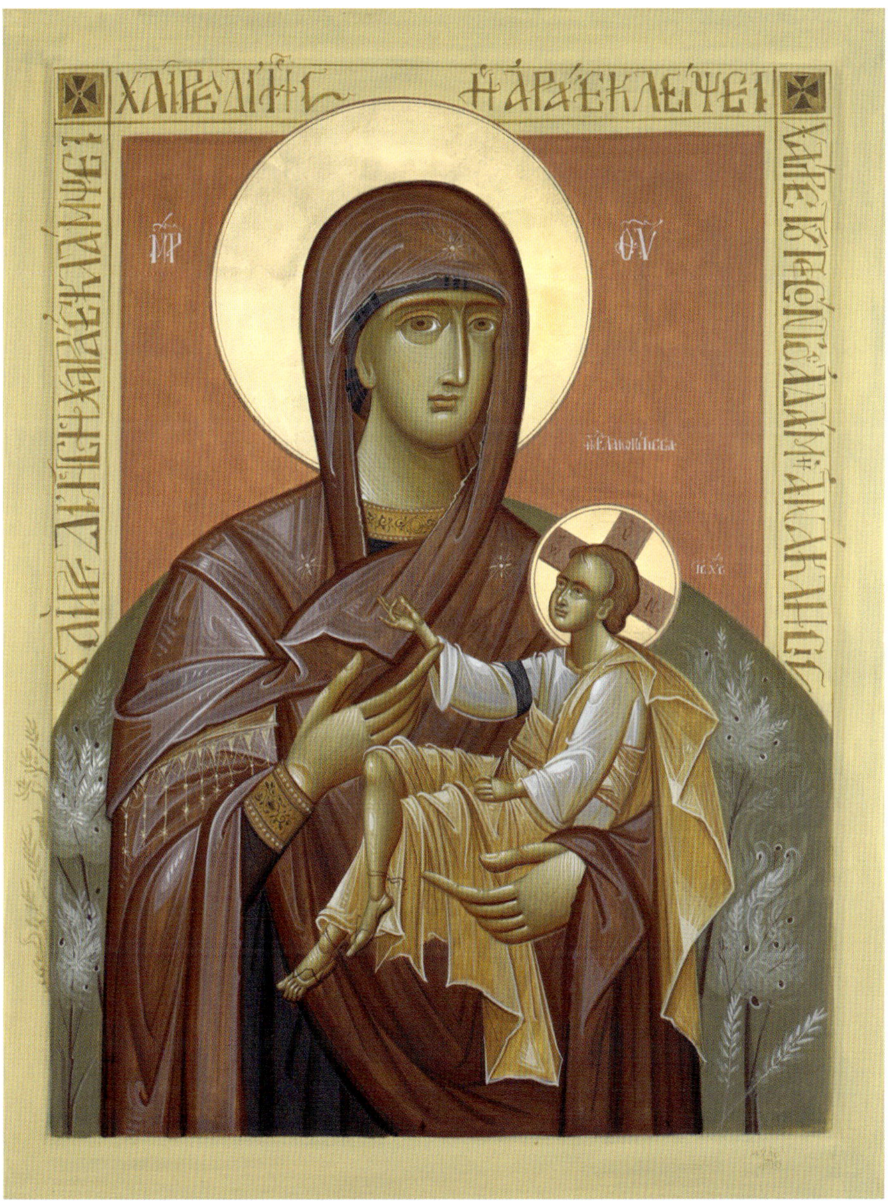

THEOTOKOS OF THE MOUNT OF OLIVES, egg tempera on wood, 60 x 90 cm

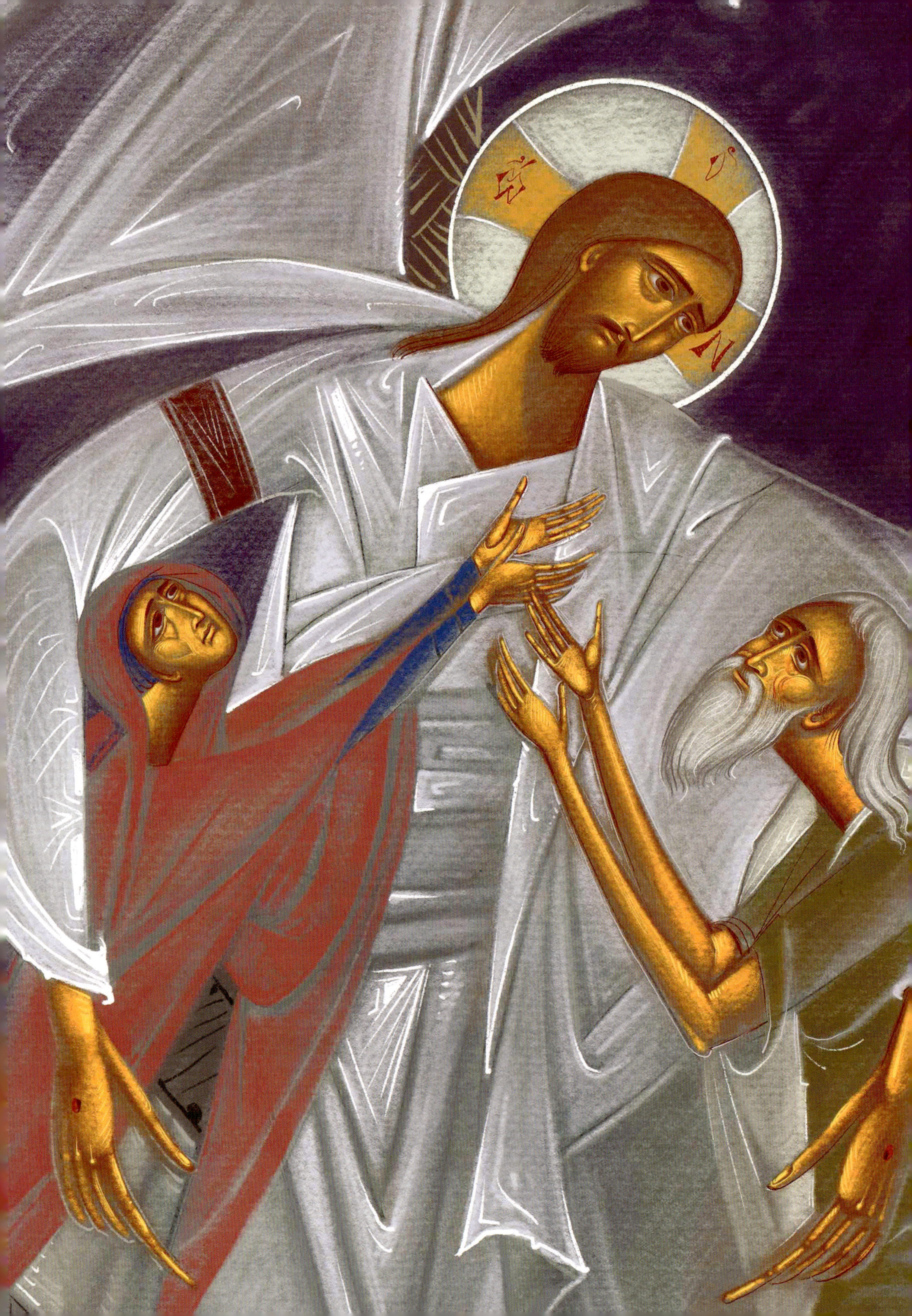

George Kordis: Form, Optimism, and the Dawn of Color

Justin Willson

George Kordis is an icon painter with a refined understanding of how color and form interrelate to create a visual narrative. In his icons color itself constitutes an element of the storyline. The contributions of color appear in the play of light and dark, with their moral connotations, and in the way color bursts onto the scene to heighten the drama or soften the jagged edge of a form. Kordis' palette is comprised of broken hues on the cooler end of the color spectrum: the painter tends to downplay strong contrasts of primary colors, preferring tints that diverge gently from one another and combine to create soft, flowing forms that complement the crisp, gently edged shapes.

The frame of an icon is a key concern in Kordis' oeuvre. Often a hand, leg, wing, billow of cloth, stone, or chain separate from the central rectangular field and enter the border, breaking the image's diegetic line. The color and form of these elements are especially important, as they bring the image out into the world, involving the sacred events in the lives of the viewer who, in turn, is invited into the fable. The unspoken idea is that the image world is theologically contiguous with the empirical world, that the possibilities presented therein are those of life beyond the pictorial surface.

Kordis' figures are historically rooted in the late Byzantine—Palaiologan—period. With their classically conceived forms, attenuated proportions, statuesque disposition, and animated gestures, the dramatic cast of sacred figures speaks to the artist's interest in an expressive style. In later Byzantium, stylistic choices became self-aware as just one artistic mode among many others available to eastern Christian workshops which had absorbed ideas from the Crusaders.

Kordis shows a historicist mentality in other ways as well. Byzantine painting is an art form of light. Greek and Slavic masters combined tempera to make the colorful garb of characters glow and shimmer, using the techniques of "assist" (geometric applications of gold leaf) and accents of white paint to pull the form out from behind the panel's surface. Over the centuries, as icons aged and their colors became diluted, these passages became diaphanous. Features of Kordis' pastel color palette speak to his consciousness of patina, combining an original emphasis on the play of transparent hues with a sensitivity to the aging processes of medieval panels. These qualities are all the more accented by an absence of gold leaf.

Color in Kordis also conveys ontological difference. On many of his vita icons, he renders the border scenes of the saint's life in a monochromatic, grisaille-like technique. Here the historical events of the saint's earthly existence are pushed to the outer edge of distinguishable pigments. The muted hues of the frame contrast with the bright, effervescent portraits at the center of the icon. Color on these panels operates hierarchically to reveal an eternal realm beyond the vicissitudes of time.

Kordis delights in visual details taken from the natural world. His representation of a curly plume of smoke wafting away from a rooftop, a wavy, striated line on a slab of marble, or a tuft of grass brushing against the fringe of a saint's cloak—these and other details fill his panels with a human element. They imbue his icons with a mild tone of delivery, casting the sacred event as a recognizable world.

Form for Kordis is a highly expressive device. While he takes lessons from twentieth-century High Modernism, which distorted Euclidean geometry to challenge academic and naturalistic pictorial conventions, his visual lexicon is ultimately Byzantine in origin and endpoint. Kordis' modeling of figural bodies, his use of gesture and the hand of blessing to convey dialogue and narrative, the disposition of figures into a compositional whole, and the fusion of portraits to vita scenes—all these interlocking relations exhibit his adherence to Byzantine types—or models—while at the same time creatively extending that tradition. Said differently, Kordis uses form to rediscover the ever-present possibility for innovation within an ancient visual canon. His icons distill an unchanging theological truth, while welcoming the viewer to look again from a fresh angle.

The brittle angularity and lyrical integrity of Kordis' figures are aimed at awakening piety. While rarely sentimental, the painter contorts bodily postures and devises compositional rhythms to unite bodies within a landscape or architectural ensemble and to underscore their mortality and frailty. More than in Byzantine icon painting, compositional lines sweep across the surface, welding together disparate sections and blending them in a syntax of pious feeling. Overall, Kordis' iconography can seem maximalist in its emotional tenor, although he avoids excesses of pathos, cautioning the viewer against reading too much humanity into the emaciated forms. Thus, he often steers clear of a fixation on scenes of brutal violence against the saints. Rather his figures are relics of a blossoming spiritual world that enters, at times, half exposed into the visible realm. These persons are artifacts of an unfolding mystery that spans epochs and comprises the shared body of the Church.

Like Byzantine artists before him Kordis is principally interested in the communicative and transformative potential of the human body: how the saintly individual serves as a site of interrelations—to the world, to the divine, to the viewer—that reveals its potency as a form of "address", or dialogue. While his icons often include successive pictorial planes, whether a brownish cityscape that recedes into the background as a patchwork of bobbing triangles and dancing lines, or an angelic choir that surrounds Christ in a glistening aureole of light, Kordis presents the body as a potent spiritual event—a temporal unfolding that leads forward into the future. In this way, the body itself acts as an agent—a character that, in a way, drives the visual plot and serves as a vortex for attributes and landscape. Thus, for instance, the desk of an iconophile theologian transforms into a swooping surface, defying gravity but providing a visual accompaniment to the author's folded leg and the loquacious swirl of the verdant ground below. The geometry of the universe configures itself around Kordis' anthropological units, endowing them with creative primacy: they are an agent-Other that stipulates the mathematics of the world.

Kordis deploys strong, almost chiseled lines to design beards, folds of cloth, craggy rocks, and the edges of buildings to inject a sculptural quality into his scenes. His predilection for open fields of murmuring pastels yields a marble-like surface that retains solidity despite the watery consistency of his tempera. Kordis' playful visual rhetoric eliminates shadow and clears the way for uninterrupted zones of color which push ornament to the fringe. His compositional strategy is to stack independent portions of an image side by side, with an integrity all their own, and then to link them together by a tell-tale gesture, wavy line, complementarity of figures or incremental increase in the shade of a nearby object. In turn the unadorned frame speaks to his minimalist attitude towards iconicity. Kordis rarely dazzles the viewer with a display of virtuosity in a painted textile, instead adorning his panels with light—bringing forth soft rays from within his scenes and inviting the eye to the dawn of colors.

Kordis tends towards visual idioms of hope and promise. Even when his subject matter is daunting or distressing (e.g., a scene of judgment, or of public shame), how he renders the human form and his choice of colors express delight and solace. In part this reflects the incredible fluidity of his architectonic structure: symmetry, a key element of Byzantine imagery, finds little place in his work. Instead, he prefers stairstep transitions between sections of his iconography, and gazes that point beyond the field of representation, to an unrepresented—and perhaps ultimately unrepresentable—beyond. These strategies convey humility, instilling Kordis' panels with quiet. They reveal the limits of an artistic performance. There is so much that can be said with the materials of painting, but there is far more that the artist cannot even begin to formulate. In this way, the icon provides a glimpse outside its own transitional window and visual repertoire, situating its actors within a vocabulary that exceeds time-bound apprehension. They are released onto the panel in a burst of optimism.

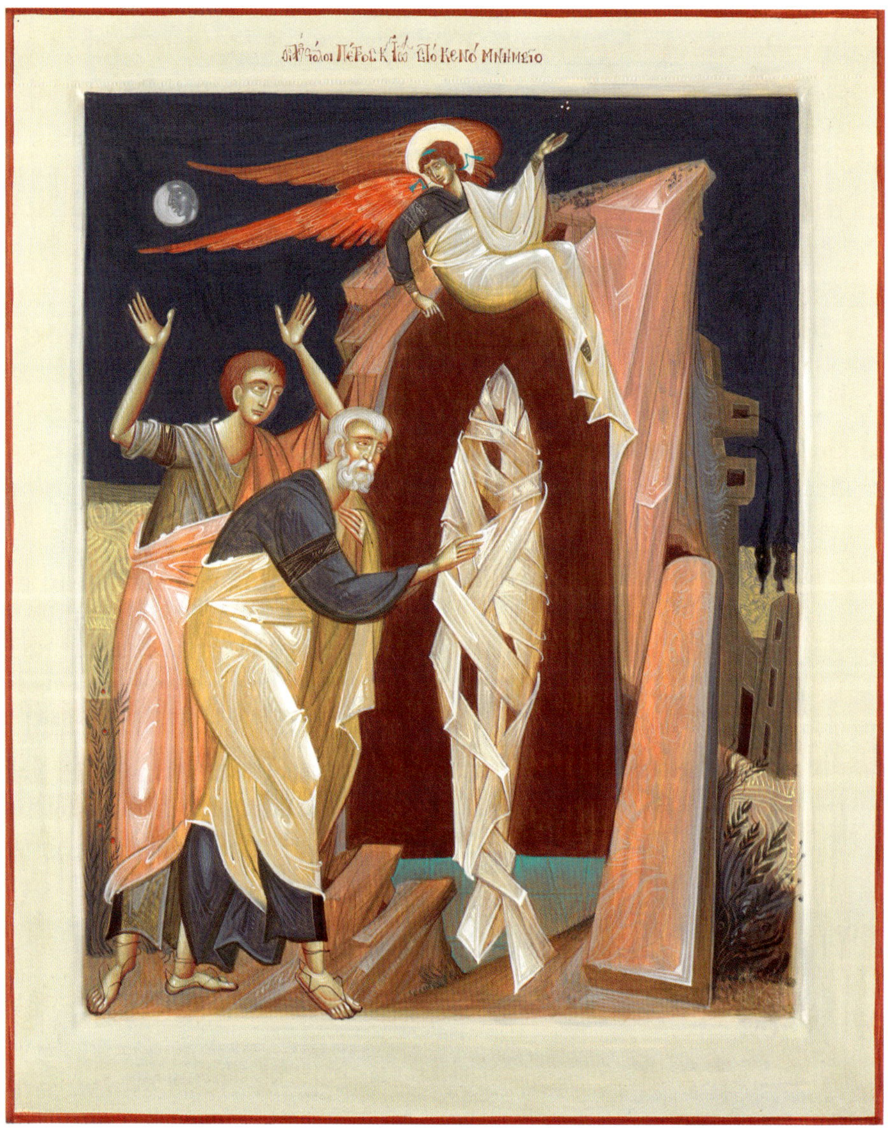

THE APOSTLES PETER AND JOHN AT THE EMPTY TOMB,
egg tempera on board, 45 x 55 cm

Kordis Painting Odyssey: Traditional and Ecumenical

Niobe Kritikou

1. Introduction

"And if it is the logos of life, it is a light that rises in the East, and if it is the world and time, it is an ever-revolving cycle."

Odysseus Elytis

George Kordis is one of the most important contemporary Greek visual artists on the international stage, whose work bridges Byzantine iconography and the principles of modernism. Wielding a visual language based in tradition, but which is simultaneously renewed through revolutionary color choices and innovations in composition, Kordis has succeeded in creating work that is deeply Hellenic, but also universal and timeless.

The revolutionary work of G. Kordis does not simply involve his development of a visual language drawn from ancient Greek and Byzantine tradition, but also his dynamic incorporation of elements of modernism and contemporary aesthetics. He is not content simply to borrow from those traditions; on the contrary, he transforms them and renegotiates them, developing new creative approaches in which the viewers themselves may actually be discerned.

In his digital paintings, he creates works of exceptional quality that combine the perfection of traditional techniques with the unlimited possibilities offered by modern technology. This undertaking is of great importance; however, it is also groundbreaking in that it proves that art can evolve without losing its authenticity and timelessness, regardless of whether it involves the use of traditional techniques or new technologies.

In his work we discern compositional reversals with strong influences from modernism, as it was expressed during the years of the avant-garde through the dynamic abstraction of Miro, the musicality of Klee's lines and, above all, the structural vision of Cézanne. The structure of Kordis' work is often both subversive and organic, built on the basis of internal axes that function as a harmonious system in the eyes of the viewer, performing an almost sacred function.

It is noteworthy, then, that for G. Kordis painting is not something static before which the viewer is a passive observer. Rather, his images invite the viewer to enter into a dialogue and to engage with his work in an experiential way that renders the viewer a participant in the essence of the work. Kordis' pictorial and conceptual world does not aim to highlight the artist himself as a prominent creative figure, but to offer an experience that conveys values, thoughts, and emotions. The artist does not stand as a conductor in front of an audience, but as a philosopher who narrates and conveys essences and concepts.

Kordis draws his themes from the human person and daily life, transforming human experiences, spiritual trials, and the fragility of the body and mind into narratives in painting. His protagonists, with their curves, long figures and outstretched necks, sweet angles and expressive gazes, bear traces of anguish, redemption, and hope. Thus, in his work the divine is not conceptualized as a metaphysical abstraction; it is a living reality as much in dialogue with Solomonic freedom as it is with the reverent spirit of Papadiamantis. These references are certainly not by chance. Kordis himself has dedicated a significant part of his artistic career to illustrating works by Greek and international writers, serving as a translator of those works into a universal language of the gaze. He is a contemporary painter, illustrator, and poet. He is a painter in the fullest sense of the word, which in Greek (*zōgraphos*) literally means "inscriber of life," because inscribed in his work are his own observations on life.

In this way, Kordis bridges the liturgical function of the icon with the secular experience of art, giving shape to an iconographic system of values in which the viewer is not a mere observer, but an integral part of the work. Thus, his legacy is not simply part of the cultural heritage of a particular place, since his work looms large somewhere between the Hellenic manner of painting and an international, universal, and ultimately eternal dimension.

2. Byzantine and Ancient Greek Influences

In the entirety of the oeuvre of George Kordis, and particularly in the series of works that appear in the book *The Joy of Cinnabar*, specific themes emerge that map out his artistic path. It is in this light that I will proceed to examine his artistic style, analyzing his expressive choices and aesthetic pursuits.

The influence of Byzantine painting on his work is evident both in his handling of line and in his approach to composition. The positioning of his figures does not follow Western practice; rather, he places them within the space as though it were a stage. By eschewing Albertian perspective, he imbues his works with a sense of theatricality. This approach derives from the post-Byzantine tradition of painting, but also from the principles of ancient Greek art, in which the movement of figures and the *contrapposto* pose impart rhythm and narrative power.

The connection of his work to ancient Greek art is not limited to a shared approach to composition and movement. Kordis has also been influenced by the philosophical understanding of form developed by the Pre-Socratic philosophers and Plato. In that tradition, ideals of beauty are not grounded in realistic renditions of subjects, but in truthful ones. In Kordis' work, the depicted figures possess an inner intensity and a sense of universality reminiscent of ancient Greek vase painting and sculpture.

It is noteworthy that linear drawing occupies a central place in Kordis' painting. His strong, almost engraved lines recall the drawings of Byzantine iconographers, as well as the lively, epigrammatic drawings of ancient Greek friezes. This approach imparts to his art an honest, handmade character that renders each work unique and authentic.

As I will demonstrate below, Kordis employs a color palette that draws from these two artistic currents: on the one hand, the luminosity and purity of the colors is reminiscent of ancient Greek frescoes and vase paintings, while on the other, the handling of light and dark tones is marked by clear references to the Byzantine technique of applying light and shadow. However, his use of colors is expressive, not realistic.

Finally, not to be overlooked is the powerful [illegible] of certain morphological elements in his work drawn from the Byzantine [illegible] of painting, including the use of outlines and Byzantine style drapery, which play a central role both in his work as an iconographer and in his freer experimentation as a visual artist. However, from within that tradition, Kordis' work is characterized by a novel expressive dynamic, where the freedom and playful spirit of the modern gaze come to reveal new pictorial means and relationships. This intensity finds expression in an enriched artistic vocabulary that is replete with signs, punctuation marks, writing, lines, colors, and various other stylistic nuances.

3. *Contrapposto* and Composition

Another characteristic feature of George Kordis' work in general is his use of *contrapposto*, which constitutes a fundamental pillar of his compositions. The principle of *contrapposto*, with its roots in ancient Greek sculpture, finds application in Kordis' painting, where it creates a sense of dynamic tension between the depicted figures. The positioning of their bodies with heads tilted, alongside other subtle alterations combine to produce a sense of liveliness and movement. The composition of Kordis' works is structured on the basis of harmonious contrasts: vertical and horizontal elements intersect, while the figures are positioned in such a way as to guide the viewer's gaze within the scene.

The playful contrasts in the positioning of the figures, as well as the structure of the compositions themselves are not merely aesthetic choices on Kordis' part, but rather constitute fundamental principles that shape his visual language. In particular, the principle of counterpoint employed in the positioning of the figures, familiar from classical tradition, here acquires a new dimension. His figures are not simply positioned in a natural manner. Rather, they become vehicles of tension in a state of transition and inner struggle, one that begins from the surface of the work and moves outward. Shifts in the centre of gravity, differentiation in the placement of limbs, and alternations between stability and dynamic movement imbue Kordis' figures with an expressive power that transcends realistic depiction.

This internal movement is not only about rendering physical balance, but also functions as an indication of internal processes. The figures seem to oscillate between stillness and levitation in a manner that signals evolution, as though in a state of constant transition. As the human body turns in different directions, it simultaneously reveals the past and the future, action and contemplation. As a rule, the figures are arranged on the basis of intersecting chiastic axes, which open up a wide angle in the direction of the audience, embracing the gaze of the viewer. Thus, even when the viewer moves in front of the work, the figures seem to follow the viewer's own movements within the space. Curved forms and angular fluctuations create an omnipresent intensity that resonates throughout the image, transforming each work into a field of internal vibration.

However, Kordis' compositions are not strictly governed by pre-determined geometrical structures, but develop in an organic way through various interactions and interconnections. The arrangement of his subjects is neither static nor symmetrical, but is determined on the basis of the work's internal rhythm, which itself depends on the relationships between colors, intersecting axes, as well as on the structural elements and organic forms that create contrasts and multiple levels in the composition, recalling a theatrical set. Each figure does not stand alone within the space but is connected to the others in the composition by means of hidden axes and interdependencies, which impart a visual coherence that runs throughout the work as a whole.

The organization of the image goes beyond the logic of static juxtaposition, following instead an internal harmony reminiscent of the orchestral arrangement of a musical composition. The depicted forms intertwine with one another, the directions of their gazes and movements serving as intelligible links that unify the space. Here, the contrasts between warm and cool colors create a gradation of color that adds to an overall sense of continuity and depth.

Although the tradition of religious iconography is often distinguished by a tendency toward symmetrical order, Kordis has stepped back from a strict adherence to this approach. His compositions maintain a functional approach to the moulding of forms and to structure, but simultaneously incorporate a new freedom, where movement, light, and the relationship between the different levels of the painting create a living and pulsating space. The figures seem to move within the icon, interacting with each other in a manner revelatory of the relationships that bind them.

His painting creates a space that transcends pure sensory experience and touches upon the deepest human need for communication, transformation, and inner revelation. For G. Kordis, art assumes its proper role when it becomes an organic and integral part of the shared life of the community in the ancient Greek sense of that word, where each individual person fulfils a specific function for the benefit of the group. So, too, each of Kordis' works in turn fulfils a function greater than the achievement of aesthetic perfection.

4. Color and Composition

One of the most recognizable aspects of G. Kordis' identity as a painter is his handling of color. In his work, color does not function as a simply decorative element, but rather as a means of expression and of creating structure. His use of color goes beyond the idea of a cohesive color palette so that each work is accorded its own autonomy and self-determination. This chromatic logic is what functions as a connecting link, uniting the artist's thought and sense of aesthetics. The colors are arranged with precision, following principles of emotion and perspective. Warm colors draw the viewer closer, intensifying a sense of intimacy, while cool shades create a sense of depth and distance. Often, a diffuse grey veil of color precedes the act of painting, functioning as a unifying background that enhances contrasts and imparts to the works a powerfully contemplative atmosphere.

His palette is characterized by lively, almost subversive color contrasts, which, though seemingly random, result in absolute harmony. Often, pastels coexist with dynamic, intense color choices, creating conditions in which color plays the role not only of rendering light, but also of intensifying the narrative. In each work, color is treated as though it were both an independent value and an organic element of the composition, with each brushstroke influencing the next, forming a multi-layered visual narrative. Thus, the autonomy of each work is realized both on the level of meaning and on that of the conceptualization of color, revealing the artist's deep knowledge and conscious use of color.

The legacy of modernism enables the artist's free use of color, which is unconstrained by the obligation to produce a faithful rendering of the natural world. Through an expressionistic and discreet distortion of visual reality, color is not a passive carrier of light, but rather an active protagonist in the narrative. Each shade seems to possess psychological characteristics, intensifying meaning and revealing shifts in emotion. The furrows of the earth are rendered in lilac, while the sky is sometimes a vibrant magenta, and at others the vibrant green of creation. Where the tension peaks, the gaze is captivated by blinding flashes and intense colors, while where the subject matter shifts to darker interior landscapes, the palette becomes subdued and is more evocative, more earthy and pastel, all following the emotional fluctuations of the subject matter from one side of the painting to the other.

This original approach to color constitutes the artist's signature, his personal imprint, which is not limited to a simple choice of shades, but constitutes an entire visual language achieved through exceptionally well-thought out color contrasts, which nevertheless give the impression of total spontaneity in accordance with the painter's aesthetics. The tonal impurities, strong contrasts, and the color range of each work are ultimately structured with architectural precision, gradually revealing a hierarchy within the work, with a conscious approach to composition that balances painterly freedom and mathematical structure. Each work seems to emerge from a quest for perfection that is conversant with contemporary art, while at the same time remaining timeless.

Of course, the structural, almost architectural perception evident in Kordis' approach to color also extends to his method of composition. The various elements of his works seem to be arranged on different planes or levels. This suggests some additional influences from the logic of Cubism, where space is broken up and restructured to express multiple spatial and temporal dimensions simultaneously. The organization of space in his work frequently recalls the compositions of Georges Braque and Pablo Picasso, as regards the self-sufficiency of colors and forms alike. At the same time, the multiple perspectives and viewing angles enable a dynamic relationship between the figures and their environment, while tonal contrasts create an intense psychological charge and are even reminiscent of the techniques of Fauvism, where color is used much more expressively than naturalistically.

The chromatic resonance of each work generates fluctuations of emotion, echoing rhythms and moods drawn from life itself. Where all is in bloom, the colors of nature shine and radiate, emitting light and warmth. When it is winter, autumn, or dusk, the shades of Kordis' colors weave atmospheres that awaken memories, whispers of the senses, and echoes of experiences with which each viewer can identify from their personal life history, ones hidden in the form of images eternally inscribed within them. This is what the artist stirs up; his images resonate with pictorial melodies, hymns, and even laments.

4.1 The Personification of Nature

Another especially notable element of Kordis' painting is the personification of nature, which enhances the power of his compositions, directly engaging with the subject matter in each case. In his works, nature does not simply function as a decorative element, but assumes human characteristics. Trees seem to lean in as if participating in the scene, the waves of the sea take on expressive forms, while the heavens often appear as if they possessed emotions. This personification of nature lends his works a poetic dimension, intensifying the connection between man and his environment.

The personification of nature in his work is manifested through the dynamics of design, his use of line, and the overall composition of landscapes. Nature is neither a static nor a neutral environment, but an active presence that interacts with the figures and the story at hand. The trees lean in as if conversing with the human figures, the waters flow with a rhythm that resembles breath, and the looming mountains are as silent observers of the action. This approach imparts to nature an anthropomorphic dimension, where the landscape is not just a space but an entity with its own existence.

The way that Kordis renders the forms of nature reinforces this sense of personification. The curves of the landscape resemble raised shoulders, the waves form spirals that recall hair blowing in the wind, while the expressive lines of flower blossoms are reminiscent of ever-changing facial expressions. The foliage either embraces the figures or stretches outward at its tips, creating an unbreakable unity between the animate and the inanimate. Rocky formations acquire expressive power and structural characteristics. The landscape becomes architectural, imparting a sense of familiarity that functions as a vehicle of communication. The organic transition from the natural world to the human element operates on the level of familiarity and communicative power, creating a dreamlike atmosphere where all elements of the work seem to breathe in the same rhythmic pattern.

Kordis' color palette also functions as a mechanism for the personification of nature. He employs emotionally charged colors that convey the mood of the landscape as though it were a human face. Deep ochres and earthy browns often convey a sense of nostalgia or contemplation, while intense blues and blue-greens suggest the world's spiritual or metaphysical dimension. In turn, the unusually vibrant colors, applied in small touches, impart felicitous elements of sharpness and clarity to the work. In this way, the environment acquires a voice of its own, expressing emotions reflected in the human figures and, by extension, in the entire subject matter of the painting.

Finally, the relationship between man and nature in Kordis' work is not one of simple coexistence, but a dialogue between equal entities. The figures often seem to belong to the setting, to extend into it or to merge with the elements of nature. This coexistence suggests a primordial unity between man and his environment, where nature is not an external element but part of human existence itself. In this way, Kordis invites us to experience nature not as something foreign, but as a part of ourselves.

4.2 The Conceptual Dimension of Color and Nature

Having analyzed Kordis' use of color as a powerful compositional tool, I will now proceed to explore the way in which color acquires a conceptual dimension in his work, precisely as in his unified approach to the personification of nature discussed above. Throughout his artistic practice, his color palette, with its attention to detail, has been marked by strong contrasts between warm and cool colors. However, in some areas, color functions far more on a conceptual level than with a view to conveying a sense of realism. This technique becomes a vehicle for spirituality, since color ultimately functions to produce emotion.

In Kordis' painting, we see that light does not come from an external light source, as is the case in the tradition of Western painting. Nor yet does he employ the technique of *chiaroscuro*, where light and dark convey a sense of volume and depth. On the contrary, here light emerges from within, radiating outward like an inner glow permeating the figures and the landscape. This approach creates the unique sense of light being autonomous, subject to an internal rhythm and enhancing the self-awareness and solemnity of the work.

As another dominant element in his compositions, line gives shape to that internal lighting. It delimits the forms without enclosing them, filling them with light and functioning as a conduit, directing the energy of the image. The forms seem to pulsate within a light that is not projected from outside, but is emitted by their very existence. Accordingly, every figure, every element of nature, even the structural parts of the works, seem to have their own luminosity, a distinct intensity that animates them and fills them with conceptual power.

In Kordis' works, the engraving of lines resemble hand movements, where the intensity and speed of the drawing lend immediacy and expressiveness to the figures. This recalls the techniques of expressionism, especially the tradition of Chinese and Japanese black ink paintings (*sumi-e*), where drawings are captured with economy of means but with maximum expressiveness. Moreover, the use of fast, fluid lines which seem to flow organically throughout the composition recalls the work of Egon Schiele, where form results from inner tension. What is achieved here is a re-staging of the very influence of nature on man and vice versa.

In nature we do not simply see the depiction of a natural landscape, but a space full of carriers of meaning. Trees, mountains, waters, hands, fish, boats, ears of wheat, cypresses, fruit, and even the sun and the moon, through their personification, design and role in the narrative of the artwork, convey concepts of birth, exaltation, hope, contribution, repentance, purification. They do not merely serve as decorative elements or as a background, but as organic parts of the whole. The elements of nature participate in the work's inner dialogue, becoming carriers of meaning and psychological states. In precisely the same way as the human figures, they emit their own light, as if carrying an inner truth independent of external factors.

This use of light and line is directly linked to devotional icons, where light does not come from a natural light source, but from the divine presence within the very bodies of the saints. However, in Kordis' works this approach does not only have a religious dimension, but also functions as a modern transcription of spirituality into art. Light becomes a means of expressing the essence of things, an element that reveals the inner world of the depicted figures, far beyond their outward existence.

The conceptual use of color enhances this inner feeling. Each shade is not just an aesthetic choice, but carries the weight of emotions and meaning. Bright colors not only express vitality, but also a sense of transcendence, while the deep, dark colors create moments of reflection and inner tension. The relationship between light, color and drawing, as well as their respective use, impart to Kordis' works a distinctive atmosphere, where the seen and unseen, the material and the spiritual, coexist in a harmonious utopia.

5. Narratives of Dream and Myth

The composition emerges from within a landscape that seems to oscillate between dream and myth. Spaces and forms intertwine in an iconography than transcends realism, opening up a field where space and time become fluid, almost spectral. Arches, bridges, and architectural structures, some reminiscent of Byzantine churches and others recalling humble dwellings, coexist with scenes of daily life imbued with a ritual dimension. The figures seem to be suspended in the air or to belong to multiple planes of reality, connecting the earthly world with the transcendent.

Like an almost primordial presence, the figure of a musician with a flute serves as a mediator between worlds, like a poetic Orpheus enchanting his audience with the sound of his music, opening up the gateway to the imaginary world. His wings extend almost underground, imparting the sense of a hidden transition, of a flight not toward the heavens but toward the interior realm of myth. Another figure seems to be passing through the liminal moment of existence, recalling Hermes the Psychopomp, the guide of souls. The object he is carrying could be a container or a lamp, as if illuminating the passage way from one level of existence to another, from the visible world to the invisible one. Just like Hermes guiding souls on their way to Hades, here, too, this figure seems to be leading the way toward a transition, suggesting an intermediate space where this life and the next intermingle.

Surrealism is present, not as an exercise in distortion, but as the natural state of the image, in which metaphor and narrative give rise to a poetic world. The connection to the tradition of Byzantine hagiography remains powerful, as the figures are dispersed throughout the space, characterized by a sense of holiness and spirituality. The lighting, especially that which illuminates the conversation at table, recalls the golden glow of mosaics, a luminous revelation that brings human experience to the threshold of the metaphysical. Sometimes the atmosphere evokes the world of Marc Chagall, where reality is never fully distinguishable from dream, but rather blends together with it in an eternal choreography of memory, desire, and signs. Myth, light and darkness, a sense of levitation between the planes of the real and the imaginary, render the narrative deeply personal but also universally accessible, as though it were a collective memory resonating throughout the centuries.

At this point, the Aristotelian conception of tragedy seems to be incorporated into this perpetual sense of motion, combining the vicissitudes of life and the inner world of the human psyche. In this reality, the hero traverses a path of knowledge and purification, enriched by the contradictions of human existence, such as the encounter of dream and myth, life and death. Through the surrealist aesthetics of the work, the field where this tragedy is played out becomes the place where the ephemeral and the eternal, the perishable and the imperishable overlap in a search and revelation, redefining the upheavals of human destiny and calling them towards purification in the mind.

6. Decorative Motifs and Rhythm

Having showcased several elements of Kordis' work that highlight the breadth of its expressive possibilities, it is also worth noting the importance of his use of decorative motifs and the element of rhythm that appear in his work. In Kordis' art, pattern and line do not simply function as decorative elements, but primarily as organic structures that lend a sense of flow, rhythm, and movement to the work. Through a multitude of repeating calligraphic strokes and delicate textures, the composition pulsates with a musicality that reveals his paintings to be ever-evolving narratives rather than static images.

His motifs, inspired both by natural and urban environments, do not constitute mere references to Greek folk tradition, but are employed in his compositions as tools to create a sense of rhythmic coherence within each work. Threads of light that branch outward like veins of marble, mountain ranges that rise like polyphonic peaks, lines that move like aerial pathways left in the wake of a flock of birds: all of these aspects engage in conversation with the depicted figures in an eternal dance of shape and rhythm.

His calligraphy is not only a matter of refinement or sophistication, but functions as a carrier of motion. His brushstrokes, fine and supple like the weft of an ancient textile, form curves and rhythms that guide the eye, while their repeating patterns create a pulse reminiscent of the uninterrupted flow of a river or of the splash of leaves in a garden touched by the evening breeze. The composition breathes, pulsates, and evolves as the eye runs across the surface of the painting.

Once again, a sense of theatrically emerges through the rhythm created by the arrangement of the figures and other elements of space and form, staging a meaningful performance. As noted above, the grouping of buildings and of the elements of nature does not follow a set geometrical arrangement, but creates an internal intonation, a hidden musical structure that dictates the overall feel of the work. Thus, it emerges that Kordis' calligraphic language is not simply an aesthetic idiom, but rather the deep, inner breath of each image, where every brushstroke weaves a pictorial tapestry of rhythm and movement. Staccato notes of light and lightning-quick flashes compose a dance of movement and sound, as do the repeating brushstrokes over the waves, mountains, and plains. Arrangements of flowers and trees form airy, varied strokes, drawing us into their gentle movements. This repetition, so fundamental to the composition of his paintings, shapes not only their rhythm, but also their musicality and sense of motion, animating an active field wherein the viewer's mind can search for the ineffable.

7. Allusions to Religion, Philosophy, Greek and International Poetry and Literature, and Folklore

Finally, having completed the above analysis of some of the basic pillars of Kordis' work, it is important to highlight some significant allusions that run throughout his work, namely, references to religion, philosophy, Greek and international poetry and literature, as well as folklore, drawing from examples from the works featured in the book *The Joy of Cinnabar*. Kordis' relationship with the Greek poetic tradition should certainly not be overlooked. The figures in his works have something of the lyricism of Solomos, the nostalgia of Cavafy, the symbolic, existential search of Seferis, and the bittersweet image-making of Papadiamantis, which Kordis shares, especially in this collection of works. Through these influences, and others such as the work of Tarkovsky, Kordis' work emerges as a kind of visual poetry that balances the past and the present, the local and the universal, the sacred and everyday life. His works incorporate deep philosophical questions as well as cultural and historical references, forming images of rich cultural interest.

At the same time, as a well-travelled artist, Kordis has enriched his artistic vocabulary with images from different cultures, architectural styles, and aesthetic endeavours from around the world. His universal conception of space, influences from Eastern and Western art, as well as his profound knowledge of the iconographic tradition impart to his works a multidimensional visual language in which the local and the universal coexist.

Referring now to a few select examples, in the *Church* series we see that the houses with curved roofs and buildings in white are reminiscent of island chapels in lonely clearings open to the sky, as is the stance of the figure of the protagonist. These images are not merely depictions of their subjects; they are poetic allegories, but also powerful dialogues between local culture and universality, celebration and lament, prayer and loss.

The compositions pulsate with a sacred geometry, the color, figures, and motion together composing an atmosphere that hovers somewhere between dream and myth, like a primordial ceremony of light. The expressionist elements coexist with soft pastel textures and colors that pulsate, sometimes like whispers and at other times like revolutionary outbursts.

This unity resembles the blossoming of spring as it unfolds freely, replete with meaning and implications, with calligraphy transformed into pure, expressive freedom. Here, Kordis' visual idiom becomes a messenger of ideas, passion, and vitality.

Upon observation of these works, one notes the presence of a sense of holiness, but also of Doric monumentality. For example, in *A Church II*, Kordis presents a figure with open arms in a posture suggesting glory or prayer. This figure resembles familiar Christian representations of visionary saints or of the crucified Christ; however, here there is no cross, but instead a lunar circle that imparts a transcendent dimension to the scene. The stance of the figure recalls depictions of Saint Francis of Assisi, known for his communication with the divine through nature.

By contrast, in his works inspired by Tarkovsky, Kordis' palette shifts, moving into the shades of twilight. The background spreads out like a veil of darkness, light diffusing ever so sparingly in a lyrical dialogue. The material world seems to be wearing out, cracking and revealing something hidden beneath the surface. The way he employs multiple layers of paint enhances the textures of the surface of the paintings. In this way, he brings to mind the techniques of fresco and of old oil paintings, as if the figures were being born from the fragments of a wall that has suffered the test of time.

The space is never concrete but constantly oscillating as in Tarkovsky's films, where the boundaries between the sacred and the mundane, the past and the present, remain in a state of flux. Kordis' figures resemble prayers as they hang in the air with a hint of faith that is never certain but always anguished. A flame waits to be lit, a miracle that never comes. This is not just a painting but a ritual ceremony of matter and light, a whisper through the dust of time with pictorial memories from carved iconostases, those sacred relics that hold images of another age.

Kordis' works also warrant comparison with the poetry of Elytis, especially the *Axion Esti* in his painting *Nostalgia*, where the central role is played by a young man with angel wings like a protective shield at his back as he raises his hands before the city that rises behind him. As in Elytis' poem, Kordis' work depicts man as part of the cosmic cycle, fixed between the sun and the earth. This is a central figure that also appears in *A Church II*, albeit in a different context in keeping with the narrative of that painting's own semantic framework.

With its three figures, *Tarkovsky. Stalker* recalls scenes from the *Iliad*, especially the lament for Hector and the tragedy of Andromache. At the same time, the drama and shadows that bathe certain works evoke a Dostoyevskian atmosphere, where guilt and redemption are central motifs.

Generally, several female figures in Kordis' oeuvre are reminiscent of the myrrhbearers of Greek laments, women who mourn, beating their hands on the earth in a clear allusion to death and its attendant rituals, while declaring a silent *memento mori* with symbols such as masks, pleading hands raised, and cypress trees that unite the soul with the heavenly realm. In folk tradition, these are symbols associated with mourning and the remembrance of death.

By contrast, in the works inspired by Papadiamantis, the figures acquire a more calligraphic grace, as though they were inscribed on an old manuscript. The figures seem almost fragile, immersed in a palette of earth tones—ochres and browns, autumnal colors that recall the rustle of dry leaves and the dull glow of afternoons on the islands of the Aegean.

Here, painting meets ethnography. The figures in these works are not everyday people, but souls treading in a liminal space, caught between life and memory, like Papadiamantis' heroes, who stand on the threshold of nature and the transcendent.

"As the earth breathed out the sweet fragrance of myrrh, as the waters whispered dreams..."

Matter attains to ritual, the vestments seem to flow, the texture of the painting is reminiscent of an old rain-soaked wall, and the light falls obliquely like the last rays of the sun on the dry stone walls and the icon paintings of abandoned churches in ruins. As in Papadiamantis, here too, our gaze does not dwell on the obvious, but searches after poetry in the tiny, imperceptible signs of decay, of the earth, of memory.

In *Love in Snow*, the figures clearly carry a mournful tone recalling scenes from icons of the Deposition of Christ or the Burial of Christ, while the female figure recalls the pieta of Western Christian tradition, or even ancient Greek kore figures depicted in mourning on funerary amphorae. Caryatids, pietas, virgins, myrrhbearers, holy women and humble mothers or widows are just a few of the images that may be discerned informing Kordis' female figures.

The existential depth of these works can be traced back to Heidegger's philosophical view of angst in combination with Plato's philosophy of the essence of form. Kordis' figures are confronted with existence, recalling the gap between life and death as analyzed by Kierkegaard. A female nude, her head held in the hands of the other figures that surround her, evokes the image of the separation of the self from one's former life, an existential alienation perhaps drawn from the thought of Sartre. These references to philosophy emerge only from the multitude of spiritual dialogues that extend from end to end in the painter's oeuvre.

Finally, in the series *Beautiful Lies*, the lie in question is not a deception, but rather a trick on the eyes. The warm tones pulsate and the compositions strangely balance themselves somewhere between expressionism and an organic charm, as if the gaze itself were expanding and contracting. Here, nature becomes vivid to the senses, as if the trees were breathing and the shadows obeying not the geometry of light but the rhythm of an inner music.

In the two works of the series *A Dream Called Montevideo*, and specifically in *People Fly to Montevideo*, the figures are released from the ground, leaving their bodies behind to float in a sky rather reminiscent of a precarious open sea. The blue of this sea is deep and intense, while the human figures display the sadness of fallen angels, as if in they were in the midst of a decision or of a fall that has not occurred. In *A Ship Called Montevideo*, the composition is more stable, more grounded, its soft colors and sweet associations echoing the warmth of companionship and the idea that travel is not only wandering, but also a refuge. A family is depicted as a united body, each member supporting the other as if the family itself were the ship carrying them.

Epilogue

In the work of George Kordis, painting becomes an act of memory and renewal, a bridge connecting the past and the present, situated between Byzantine holiness, ancient Greek harmony, and the continual experimentation of modernism. Whether moving in earthly realities or floating in dreamlike landscapes, his figures carry the weight of an ever-evolving narrative tradition.

The various series—from the transformations of nature of *Beautiful Lies*, and from the contemplative anguish of Tarkovsky to the spirituality of Papadiamantis, from voyages to Montevideo to the more familiar scenes of homes lost or reclaimed—all converge in a deep reflection on humanity and its place in the world. Kordis' visual art emerges as a refined conversation with religion, philosophy, poetry, and folklore, not simply borrowing formal elements from them, but treating them as fundamental sources of inspiration.

However, beyond his rich narrative arsenal, Kordis' work also possesses a uniqueness that renders it universal. It is not limited by the boundaries of his own national tradition, but rather expands that tradition, renewing and integrating it into the great artistic heritage of the world as a whole. His painting is a window to the eternal, where images do not only have stories to tell, but also reveal fundamental truths about human existence. With his particular visual language, Kordis offers us a new perspective on 21st century art, enriching not only modern Greek painting, but also the evolution of expression in visual art on an international scale. Through his paintings, Kordis reminds us that when images are the carriers of truth and substance, they can become a point of encounter between cultures, as well as markers of memory and spiritual exaltation, a legacy for the future, and a ritual service performed for the benefit of society.

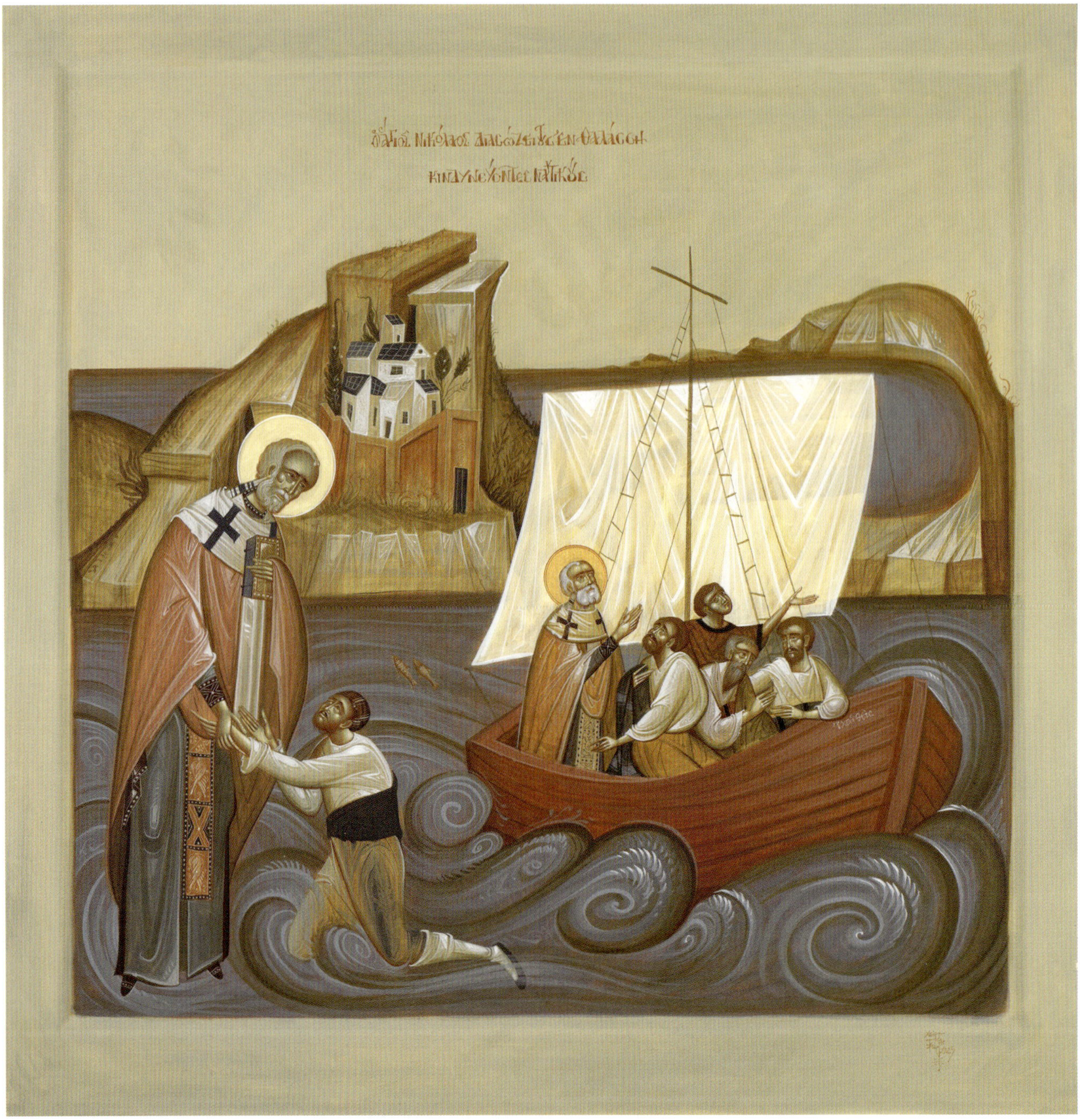

SAINT NICHOLAS SAVING THE SAILORS,
egg tempera on wood, 80 x 80 cm

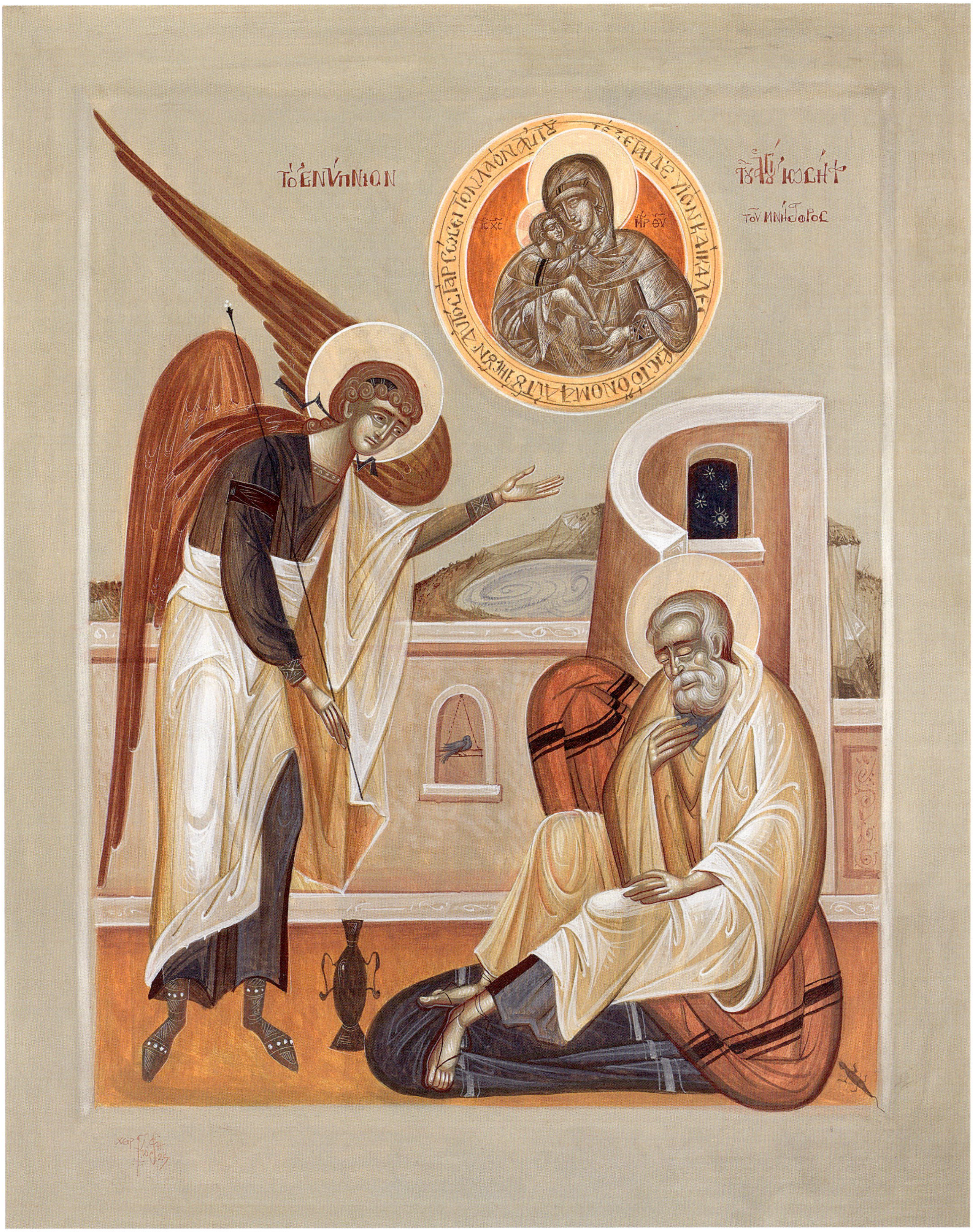

THE DREAM OF ST JOSEF,
egg tempera on board, 45 x 55 cm

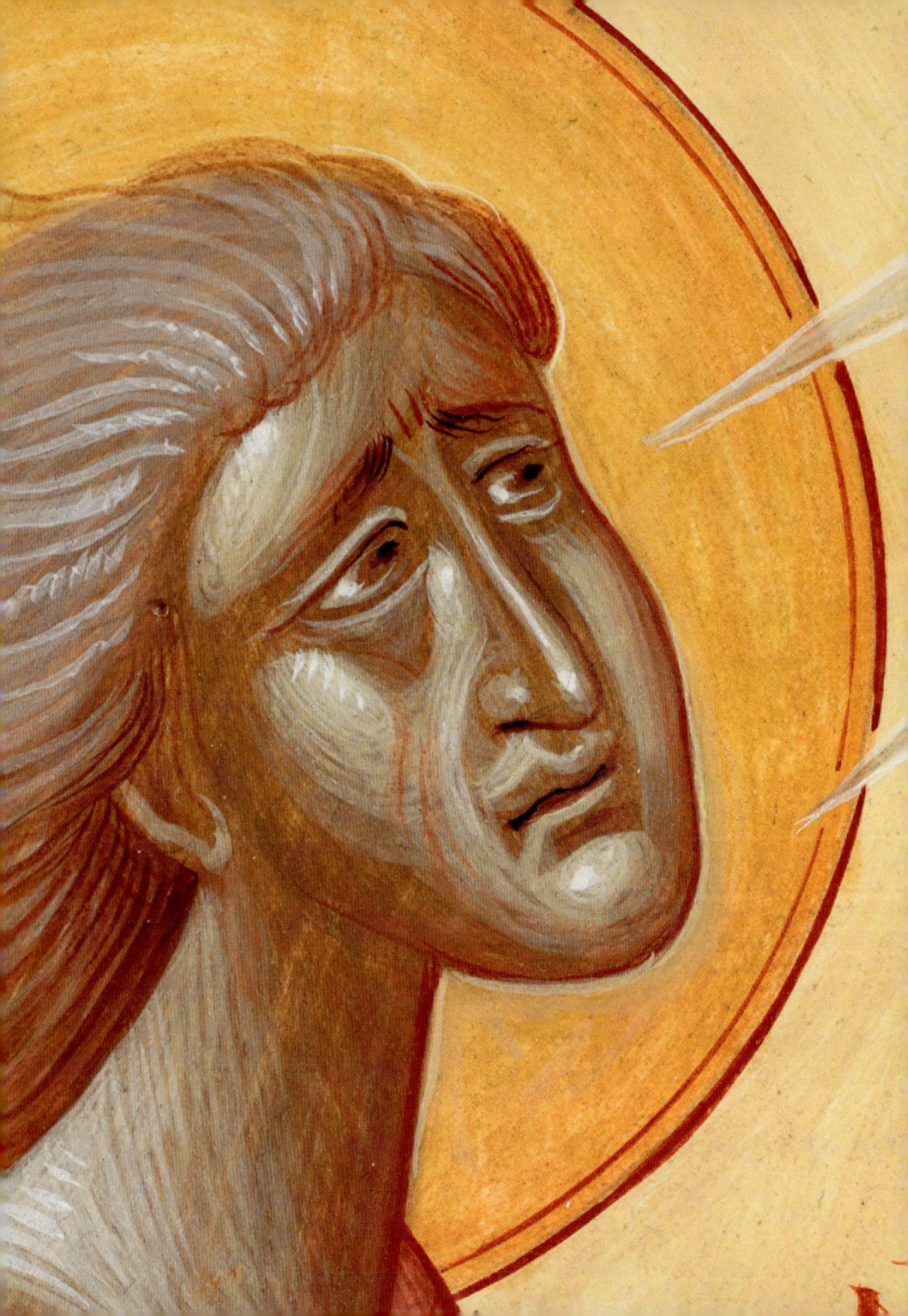

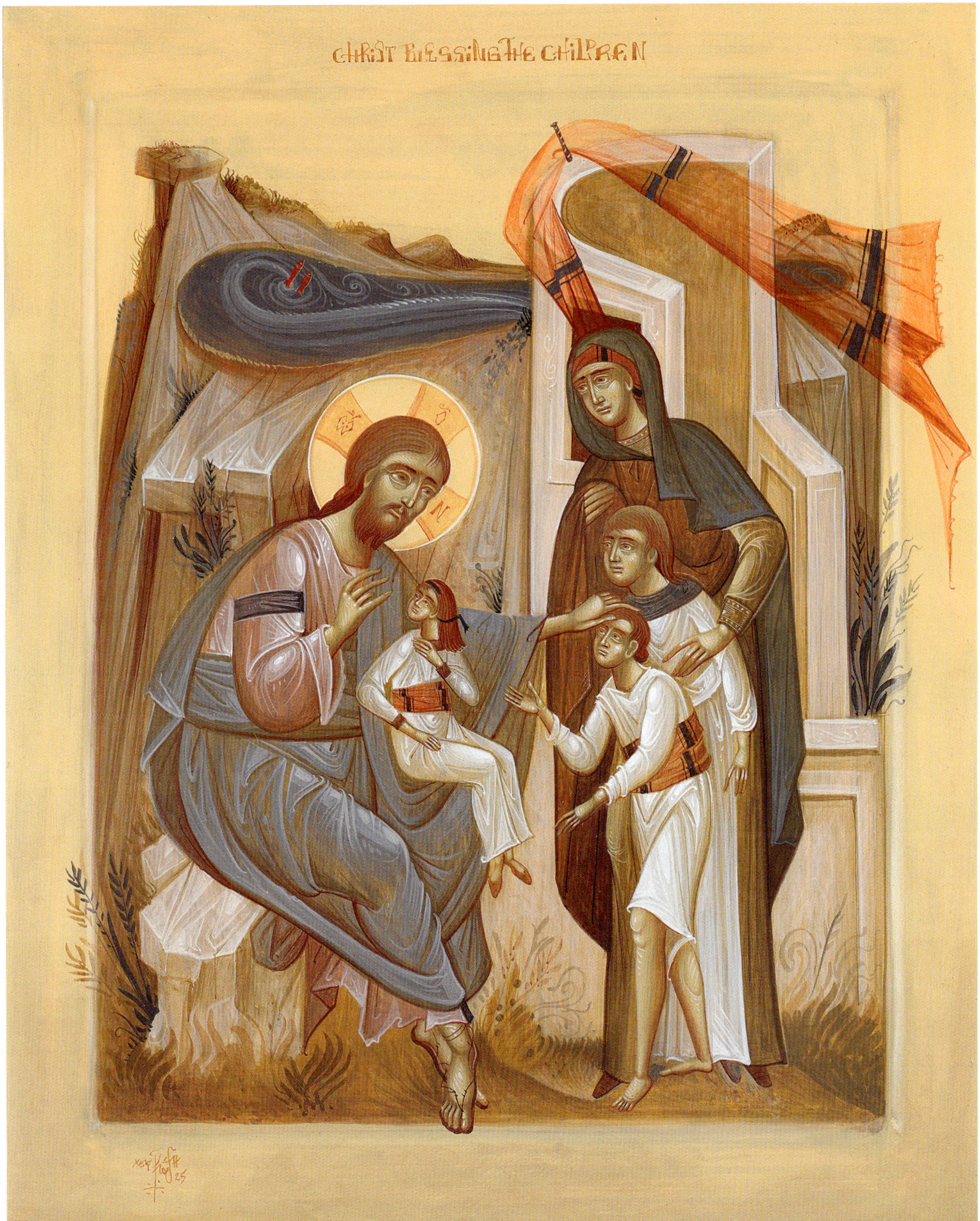

CHRIST BLESSING THE CHILDREN,
egg tempera on board, 45 x 55 cm

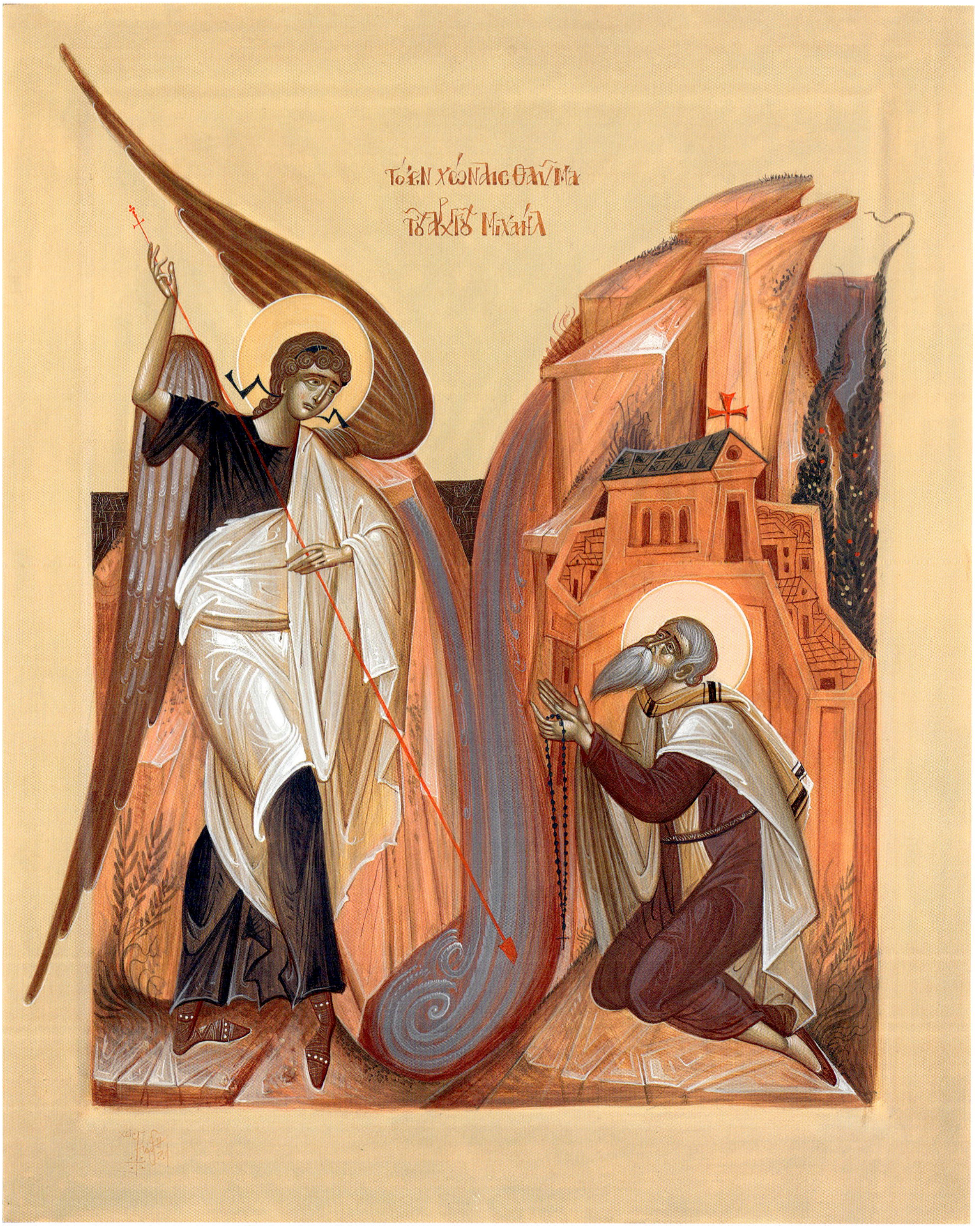

THE MIRACLE OF ARCHANGEL MICHAEL AT CHONAI,
egg tempera on board, 45 x 55 cm

TRANSFIGURATIO
egg tempera on board, 45 x 55

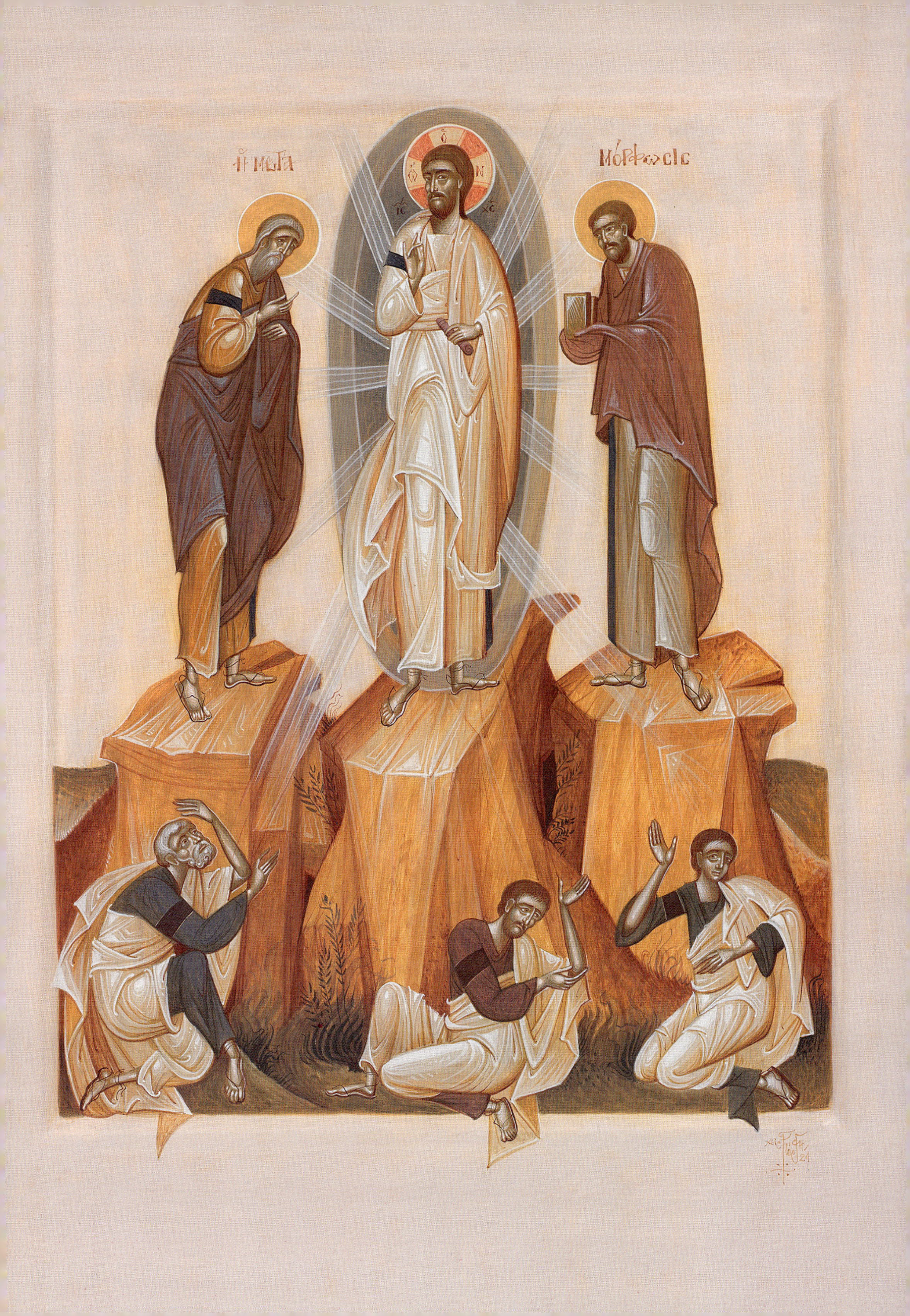
Ἡ ΜΕΤΑ
ΜΟΡΦΩΣΙΣ
Ὁ ὬΝ
ΙΣ ΧΣ

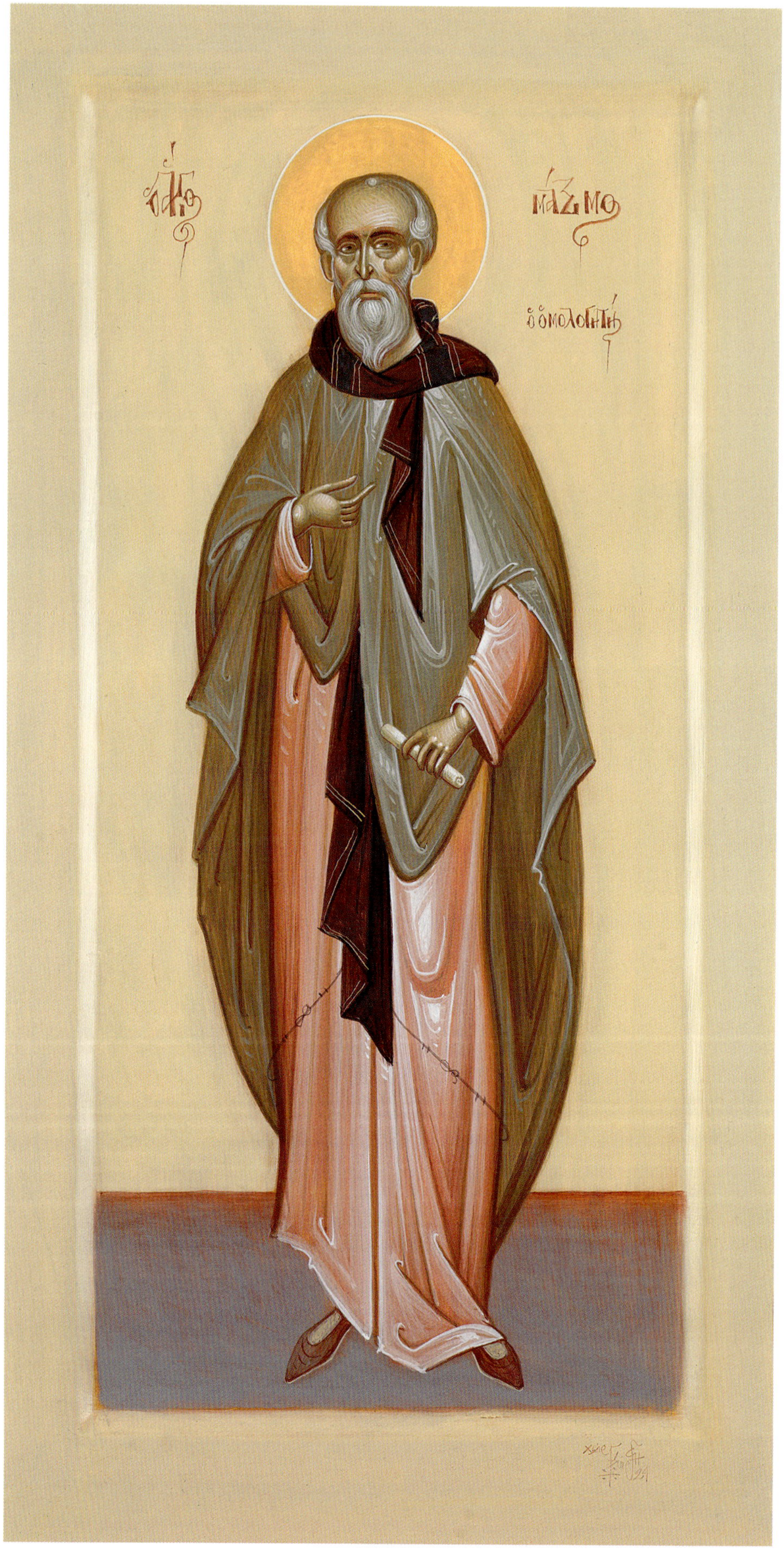

ST MAXIM THE CONFESSOR,
egg tempera on board,
27 x 54 cm

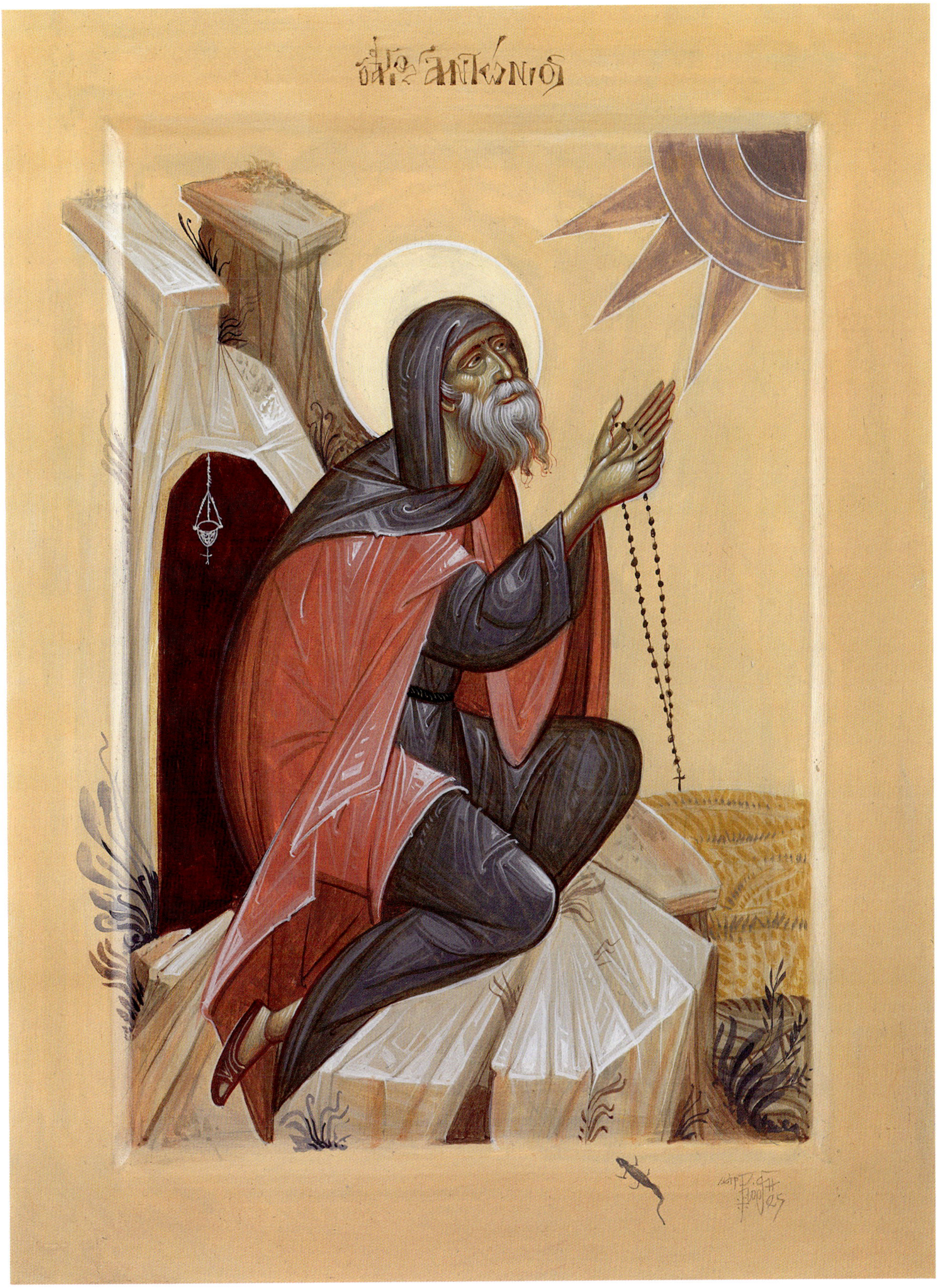

ST ANTHONY THE GREAT,
egg tempera on board, 40 x 30 cm

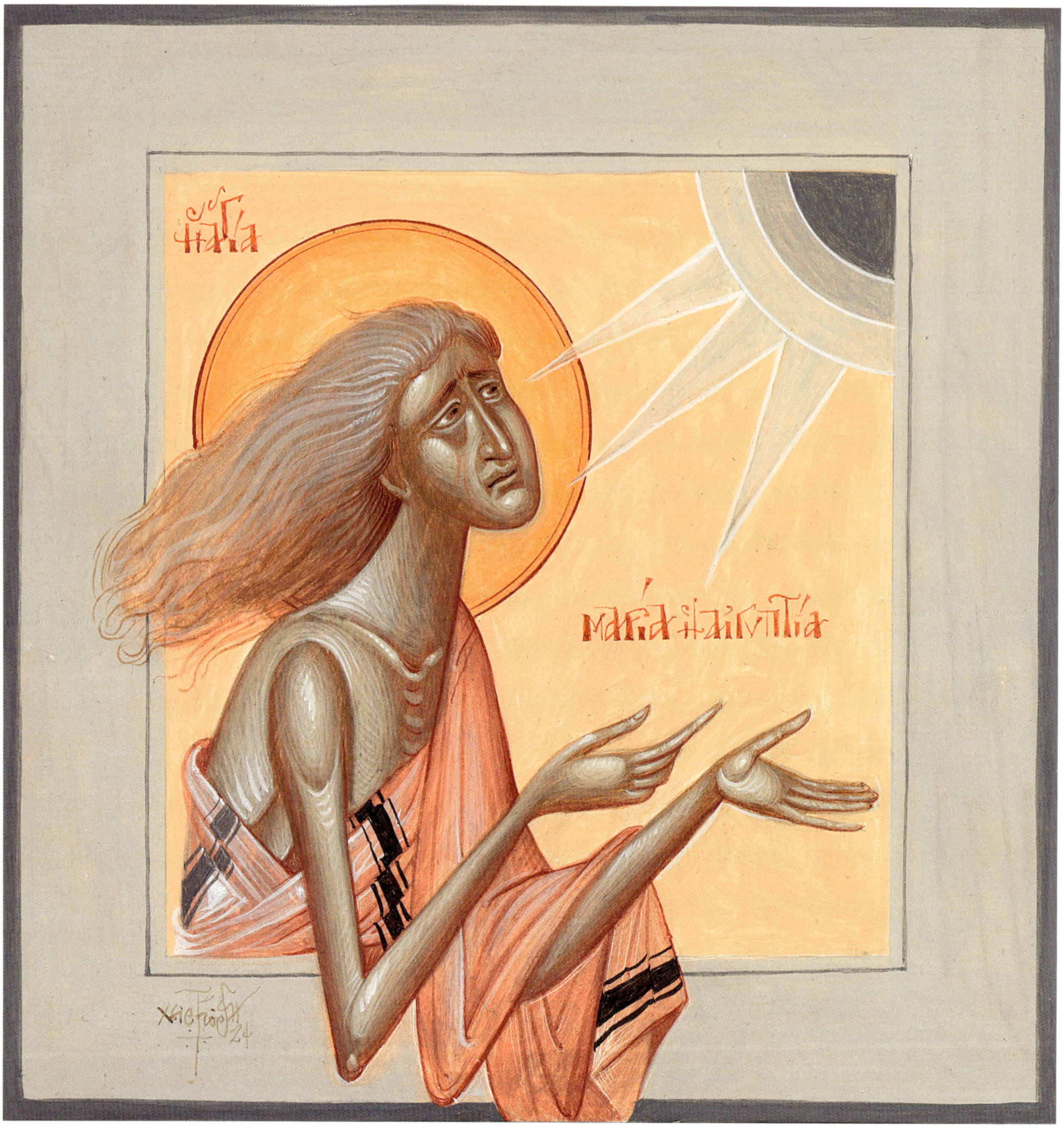

ST. MARY OF EGYPT,
egg tempera on board, 25 x 25 cm

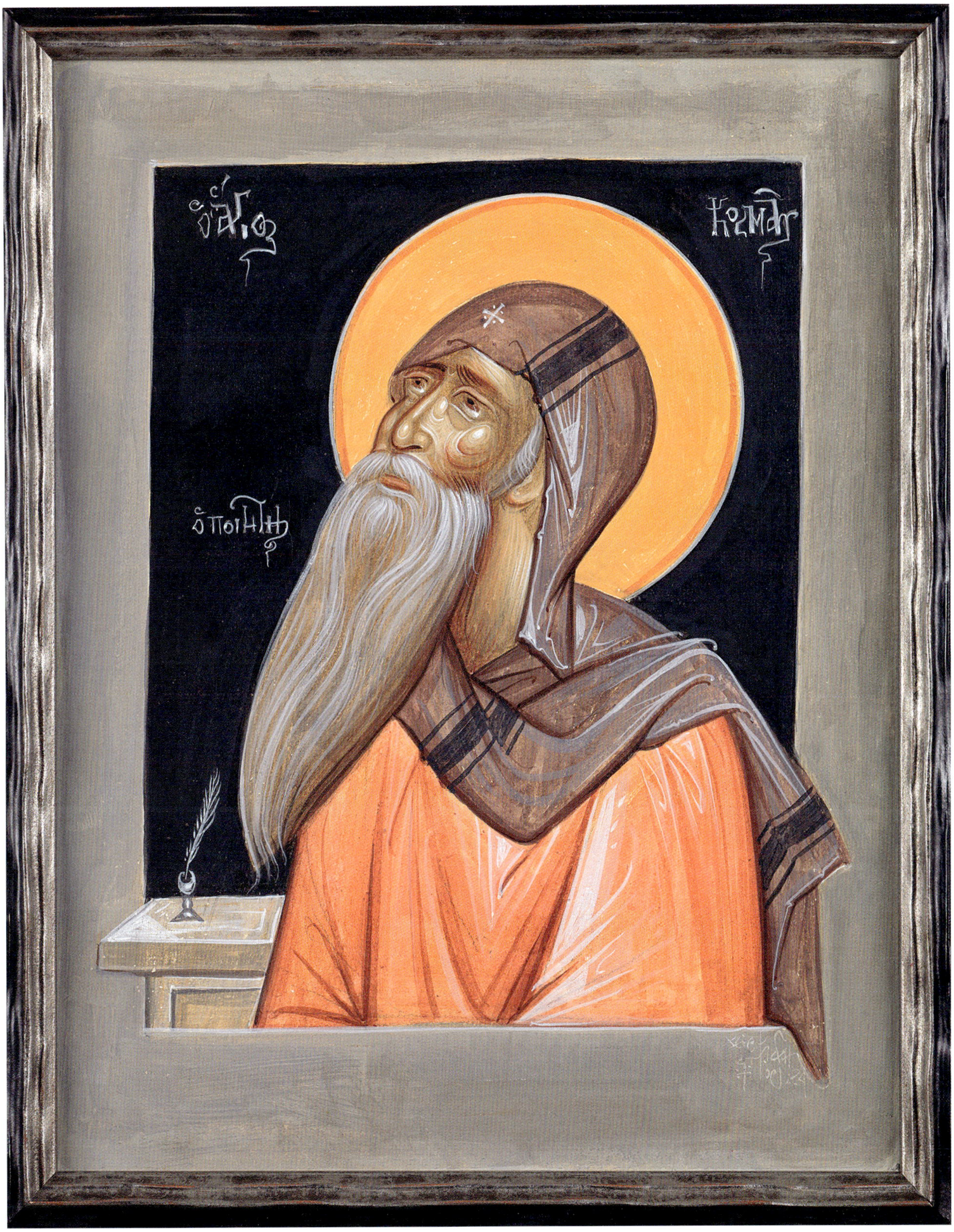

ST. COSMAS THE POET,
casein on board, 25 x 35 cm

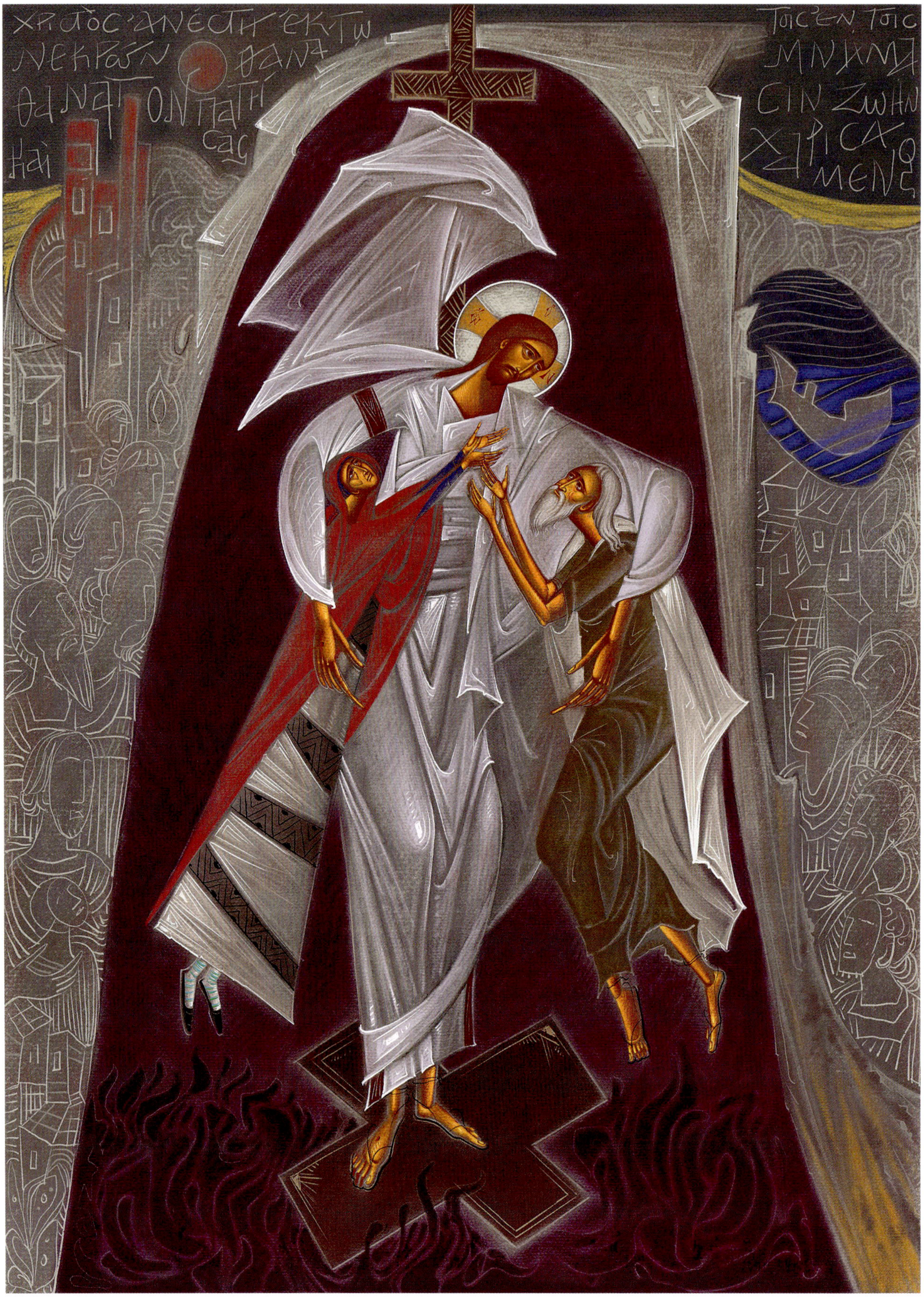

ANASTASIS,
digital painting, 80 x 160 cm

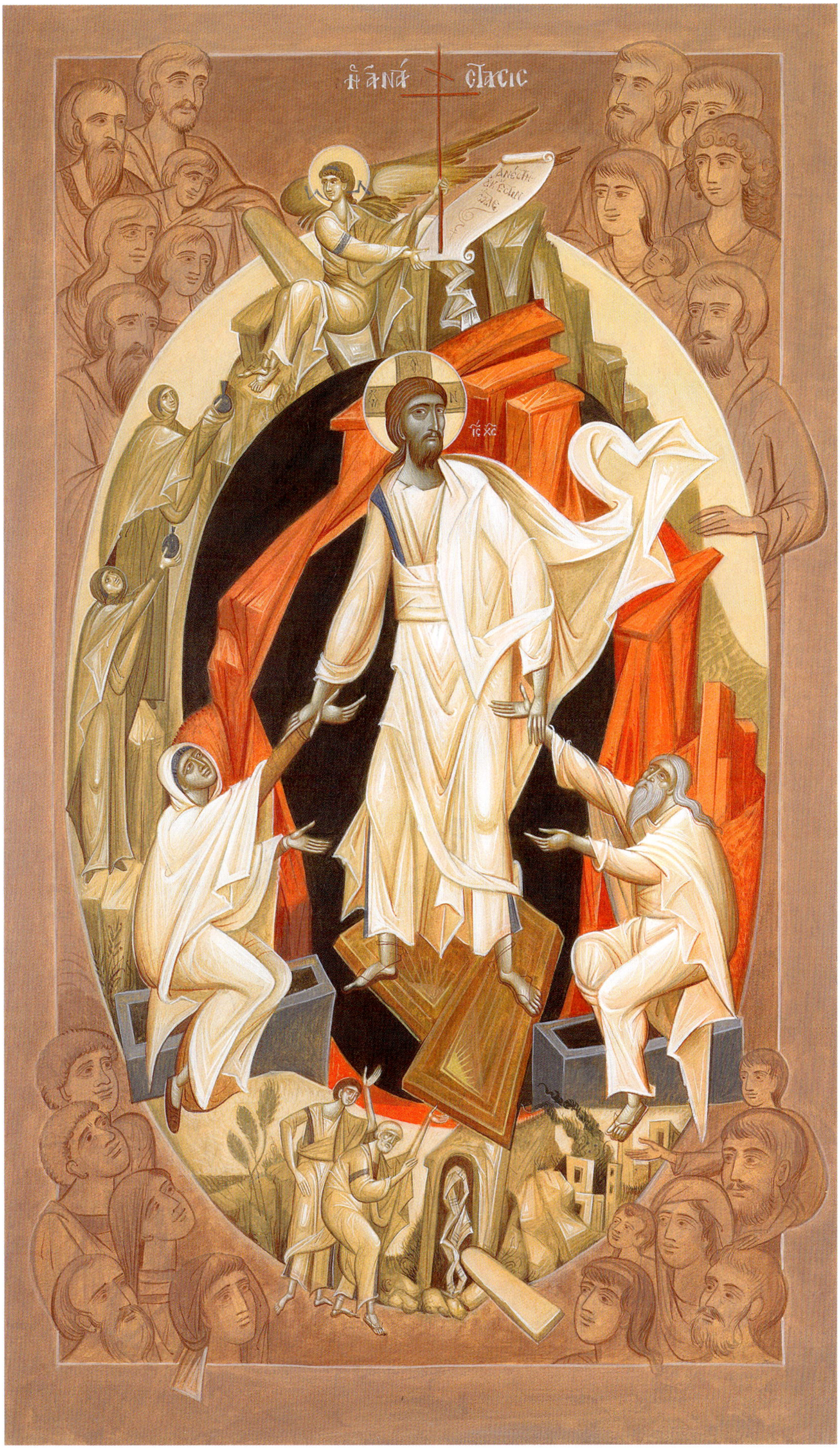

ANASTASIS,
mixed media on board,
60 x 90 cm

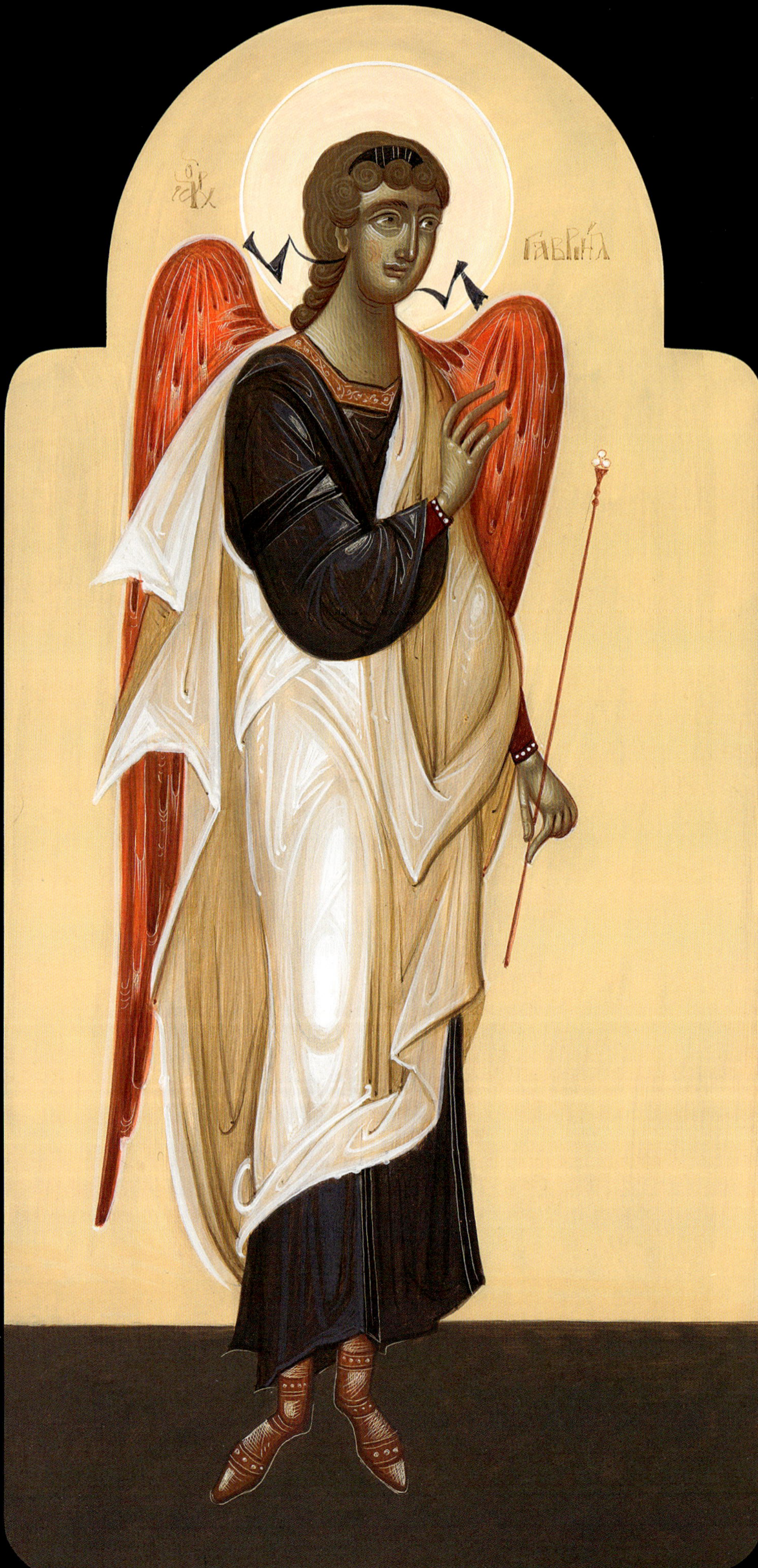
ГАВРИИЛ

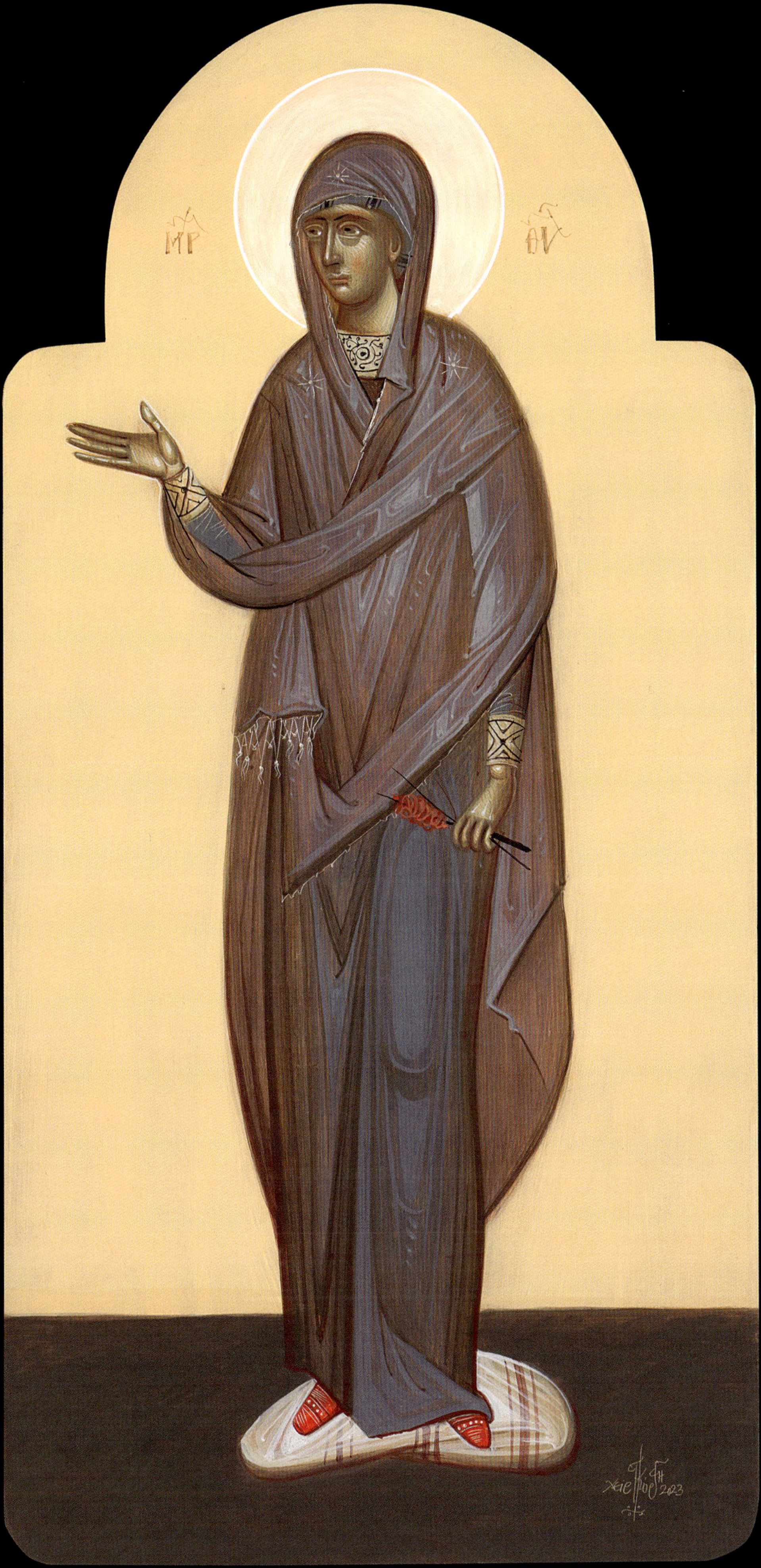

ANNUNCIATION,
egg tempera on board,
20 x 31 cm x 2

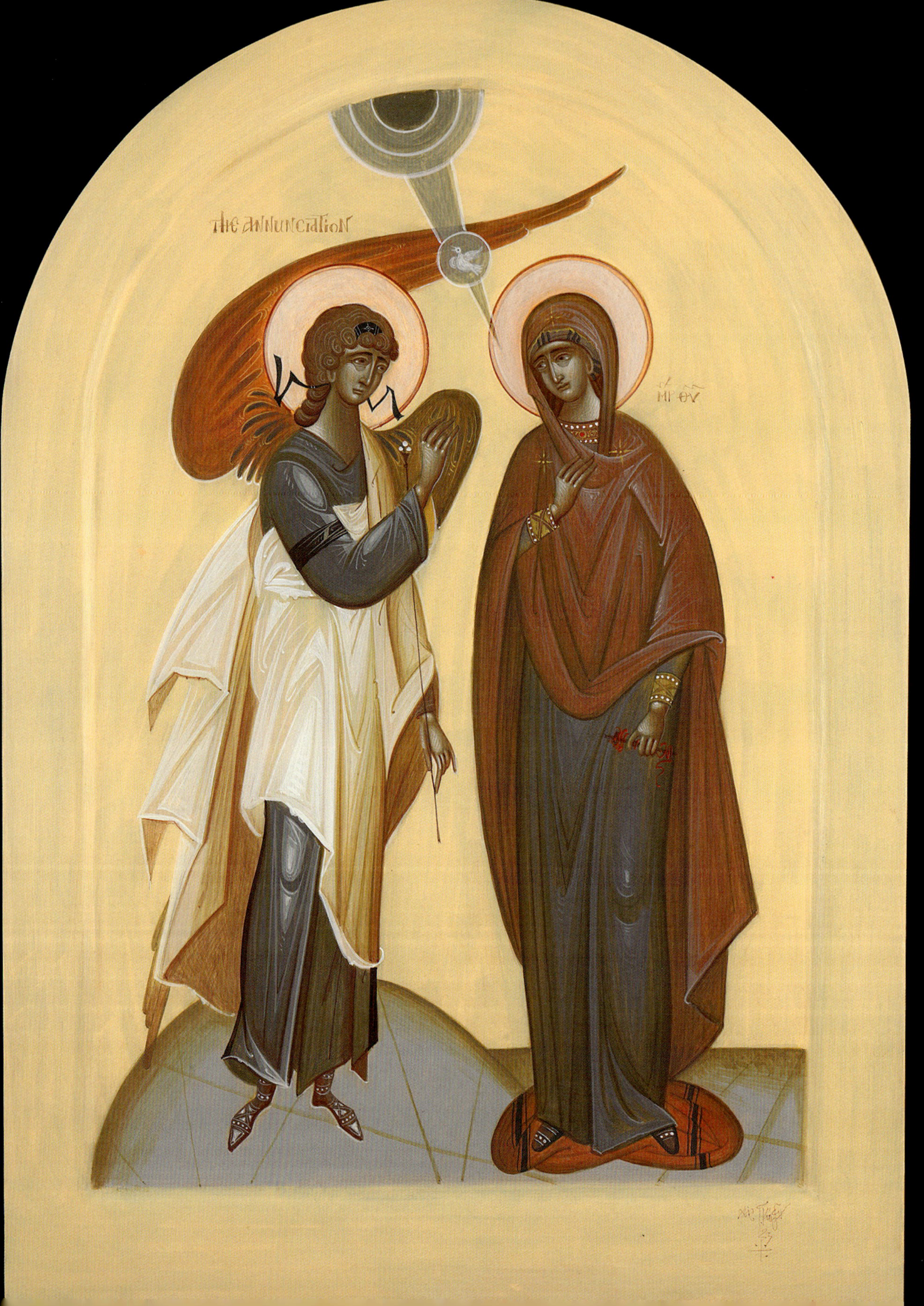
THE ANNUNCIATION
ΜΡ ΘΥ

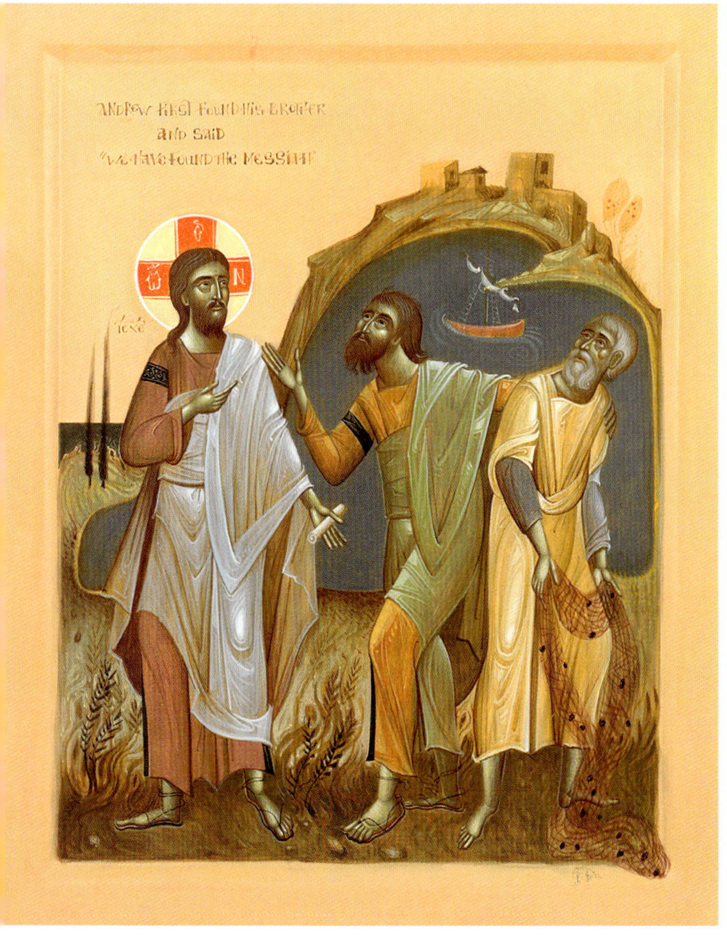

FOLLOW ME, egg tempera on board,
40 x 50 cm

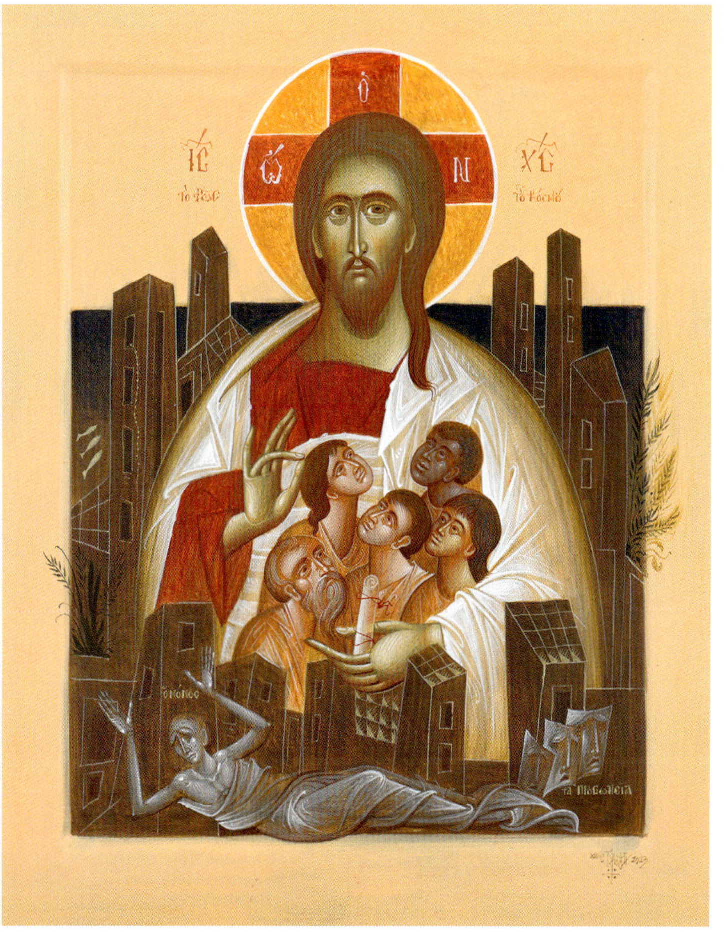

CHIRIST, THE LIFE OF THE WORLD,
egg tempera on board, 45 x 55 cm

NNUNCIATION,
g tempera on board,
) x 70 cm

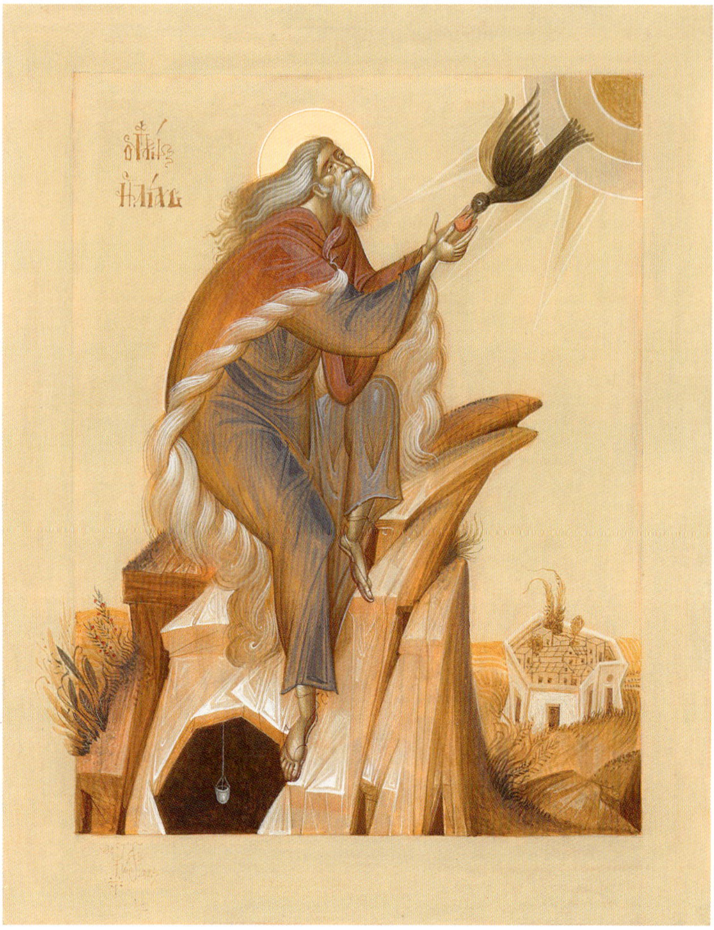

PROPHET ELIAS,
egg tempera on board,
40 x 50 cm

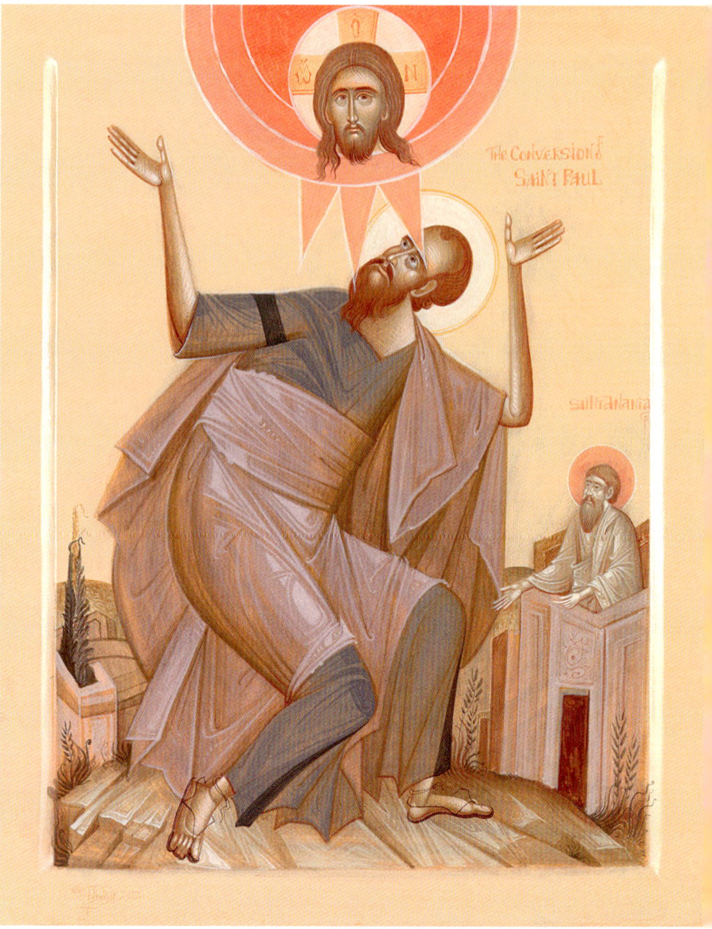

ON THE ROAD TO DAMASCUS,
egg tempera on wood,
40 x 50 cm

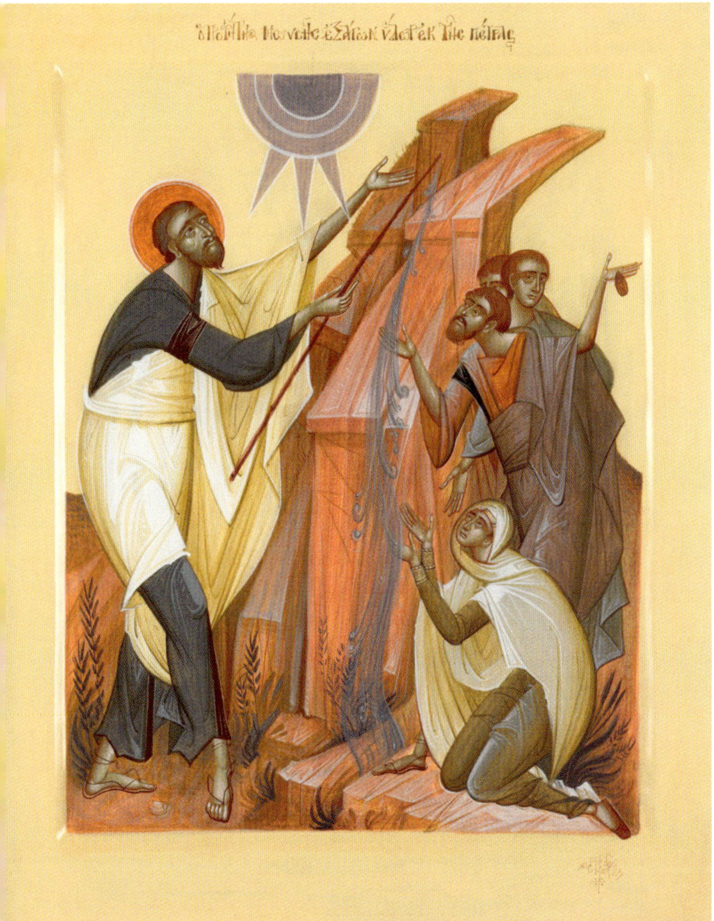

PROPHET MOSES BRINGING WATER OUT OF THE ROCK,
egg tempera on board,
45 x 55 cm

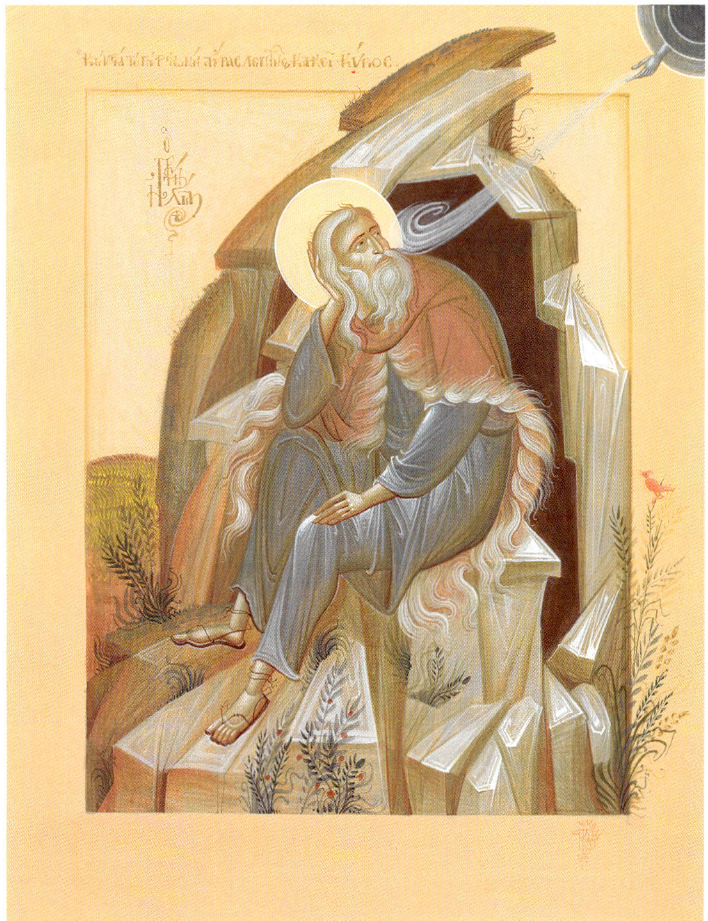

PROPHET ELIAS,
egg tempera on board,
40 x 50 cm

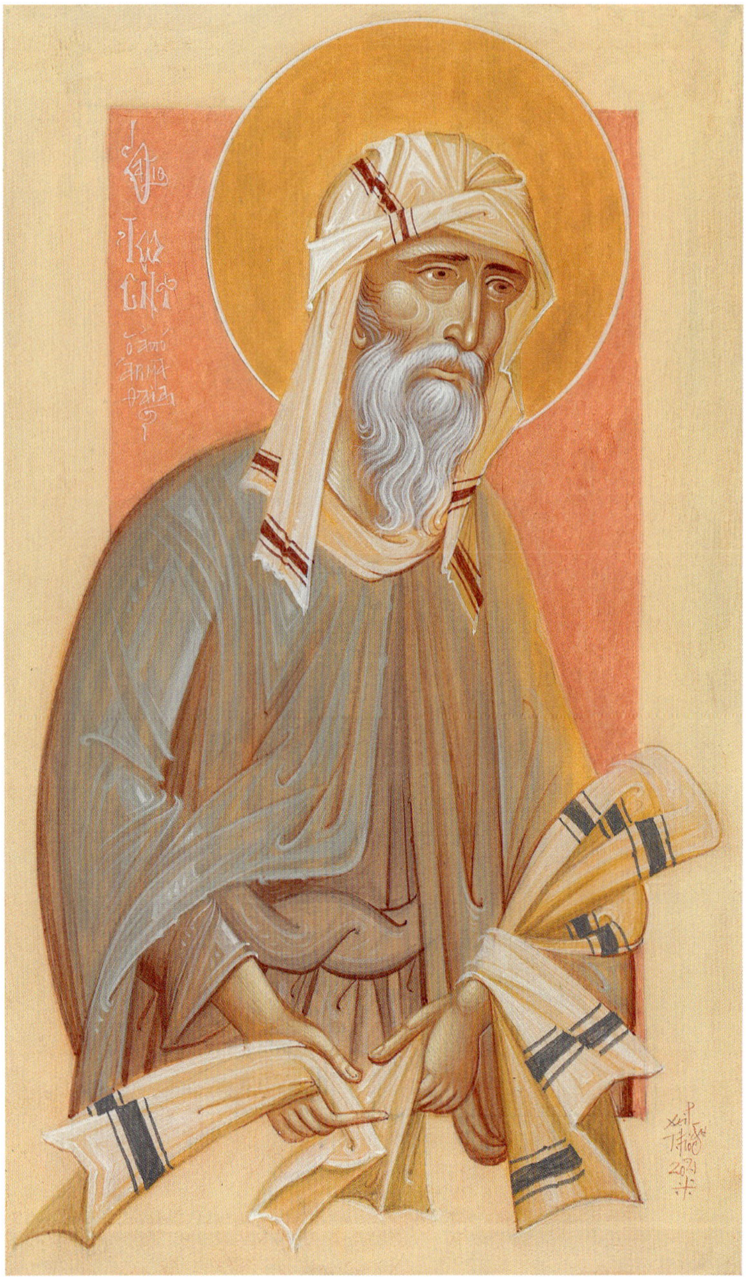

SAINT JOSEPH OF ARIMATHEA,
egg tempera on board,
25 x 37 cm

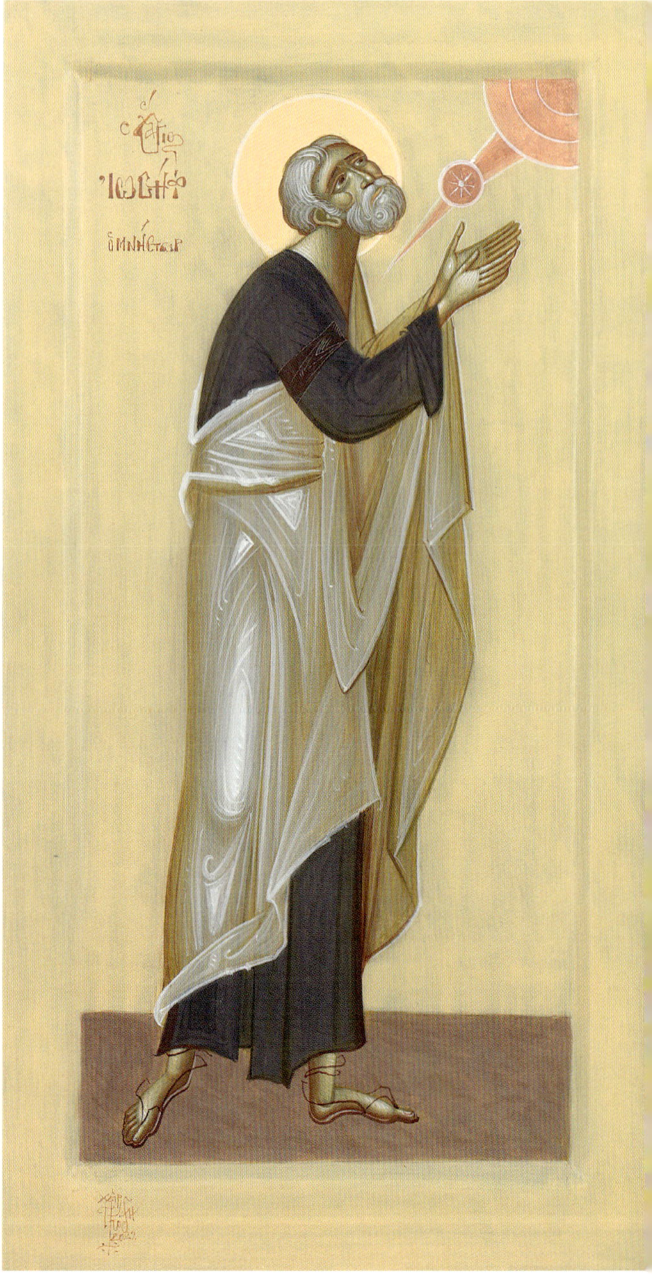

SAINT JOSEPH,
egg tempera on wood,
27 x 54 cm

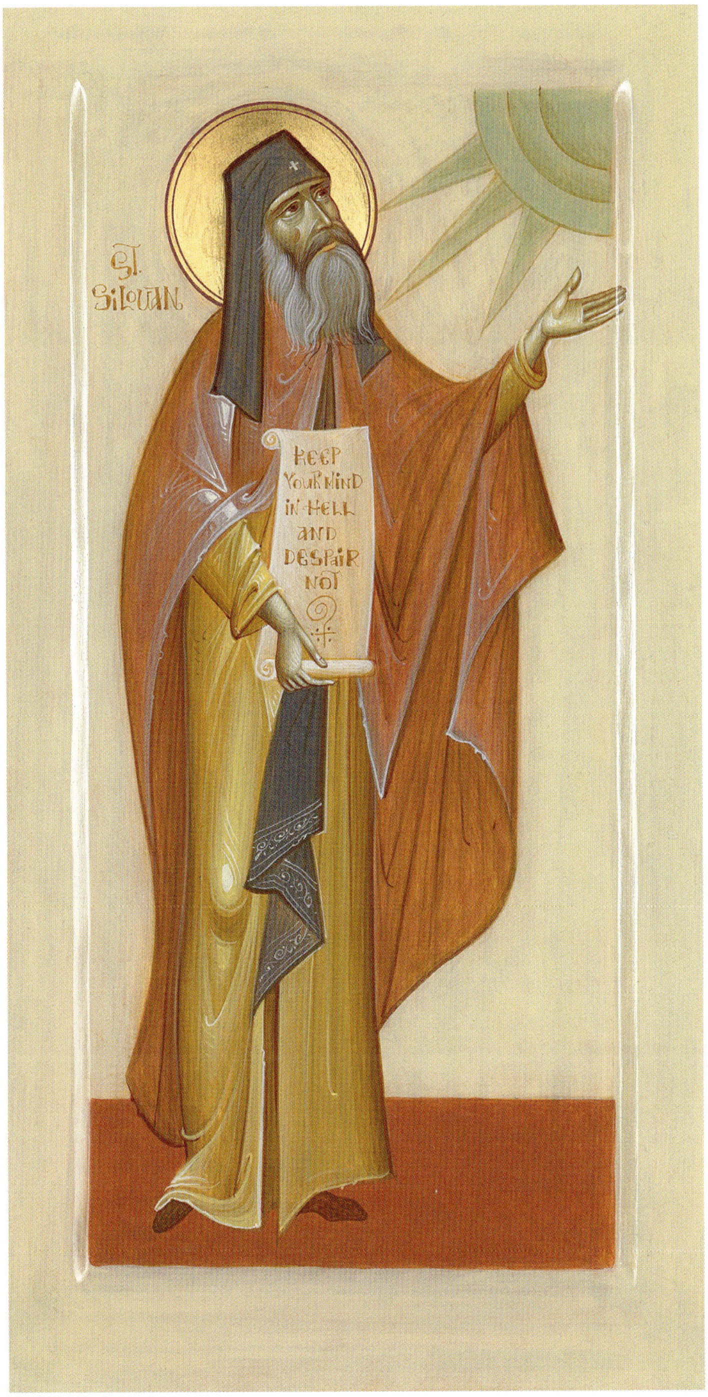

SAINT SILOUAN THE ATHONITE,
egg tempera on board,
27 x 54 cm

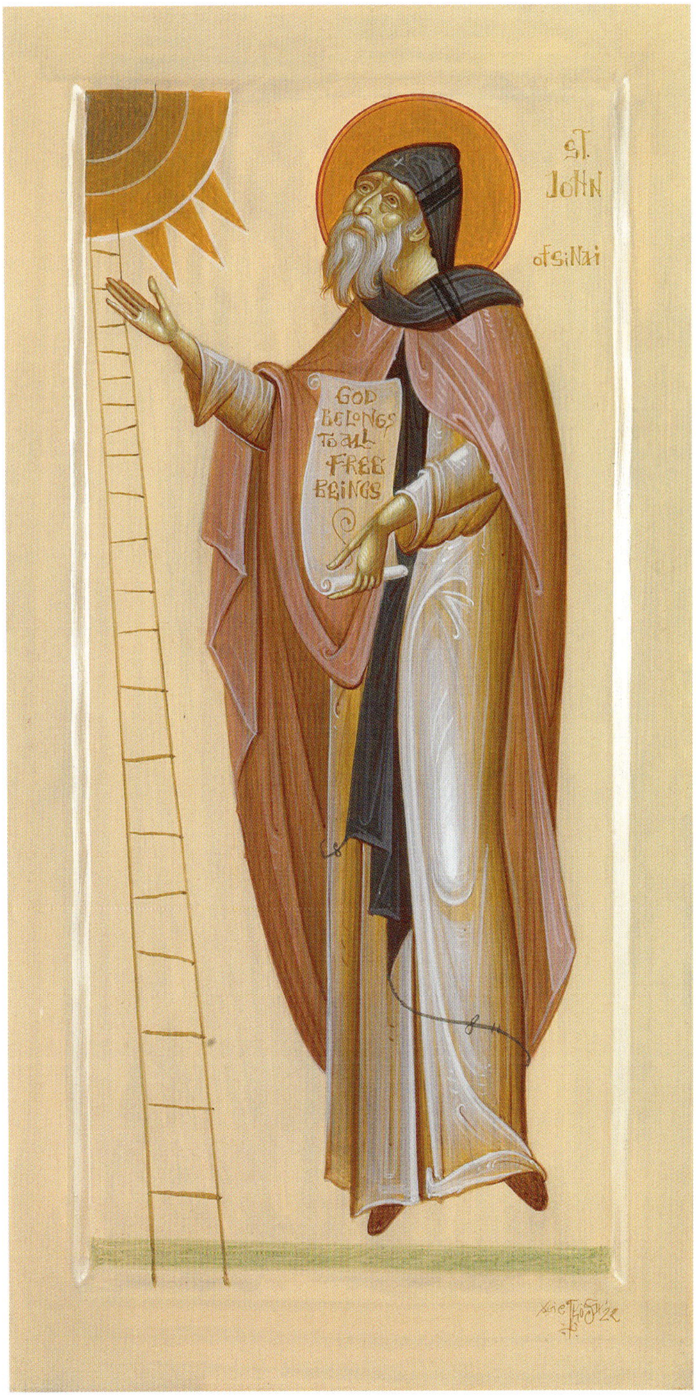

SAINT JOHN OF SINAI,
egg tempera on board,
27 x 54 cm

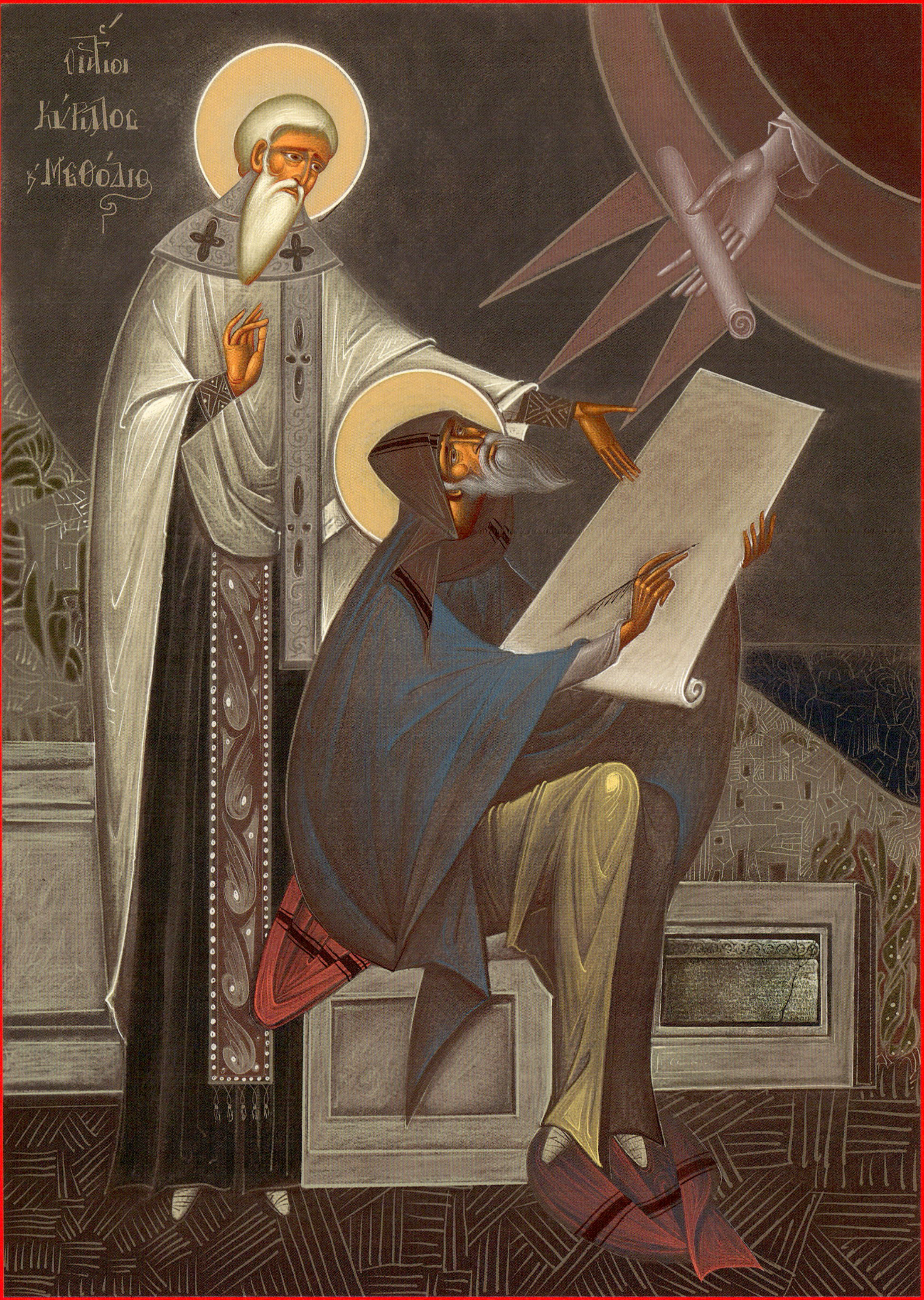

SAINTS KYRILLOS AND METHODIOS,
digital painting, 80 x 160 cm

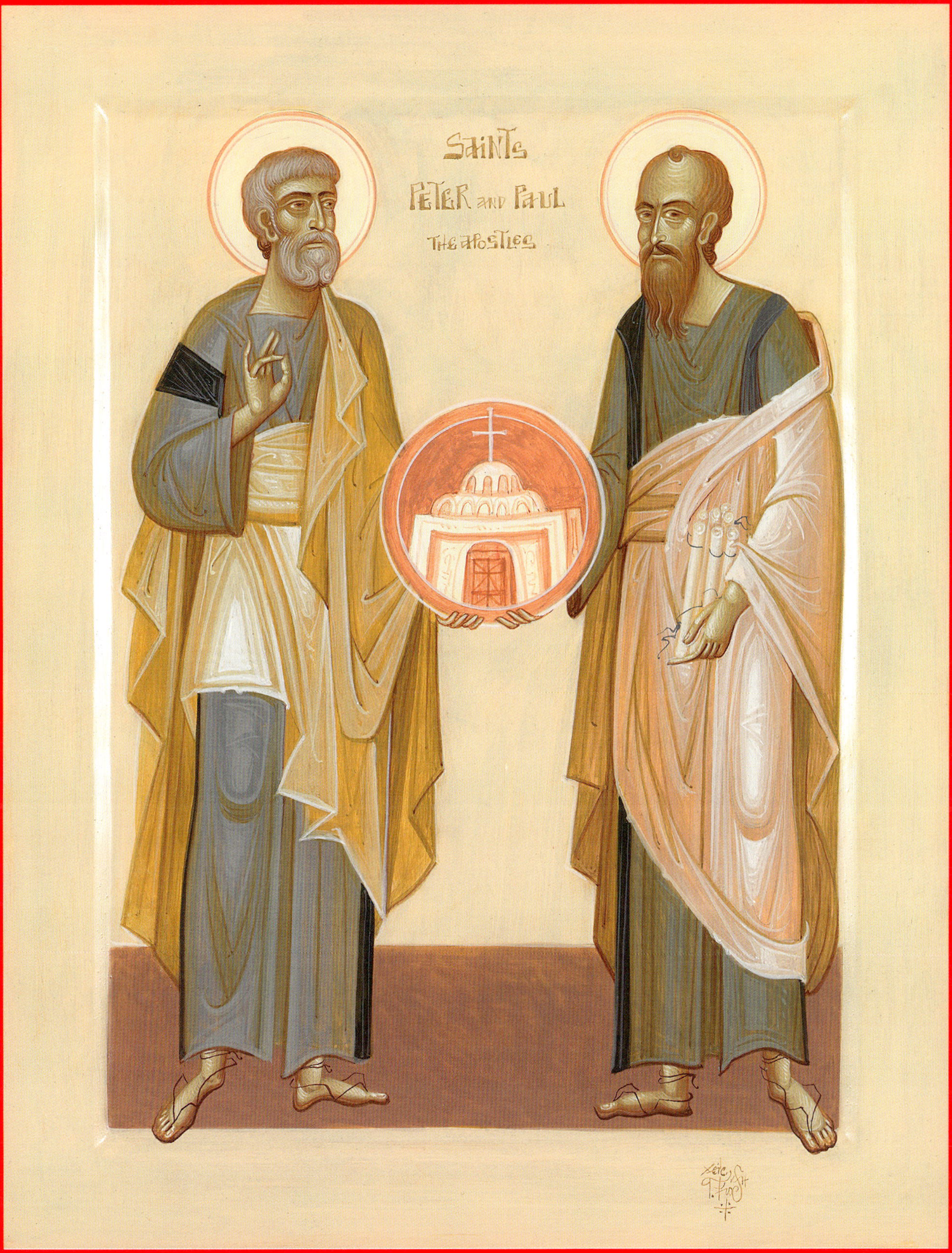

SAINTS PETER AND PAUL,
egg tempera on wood, 40 x 50 cm

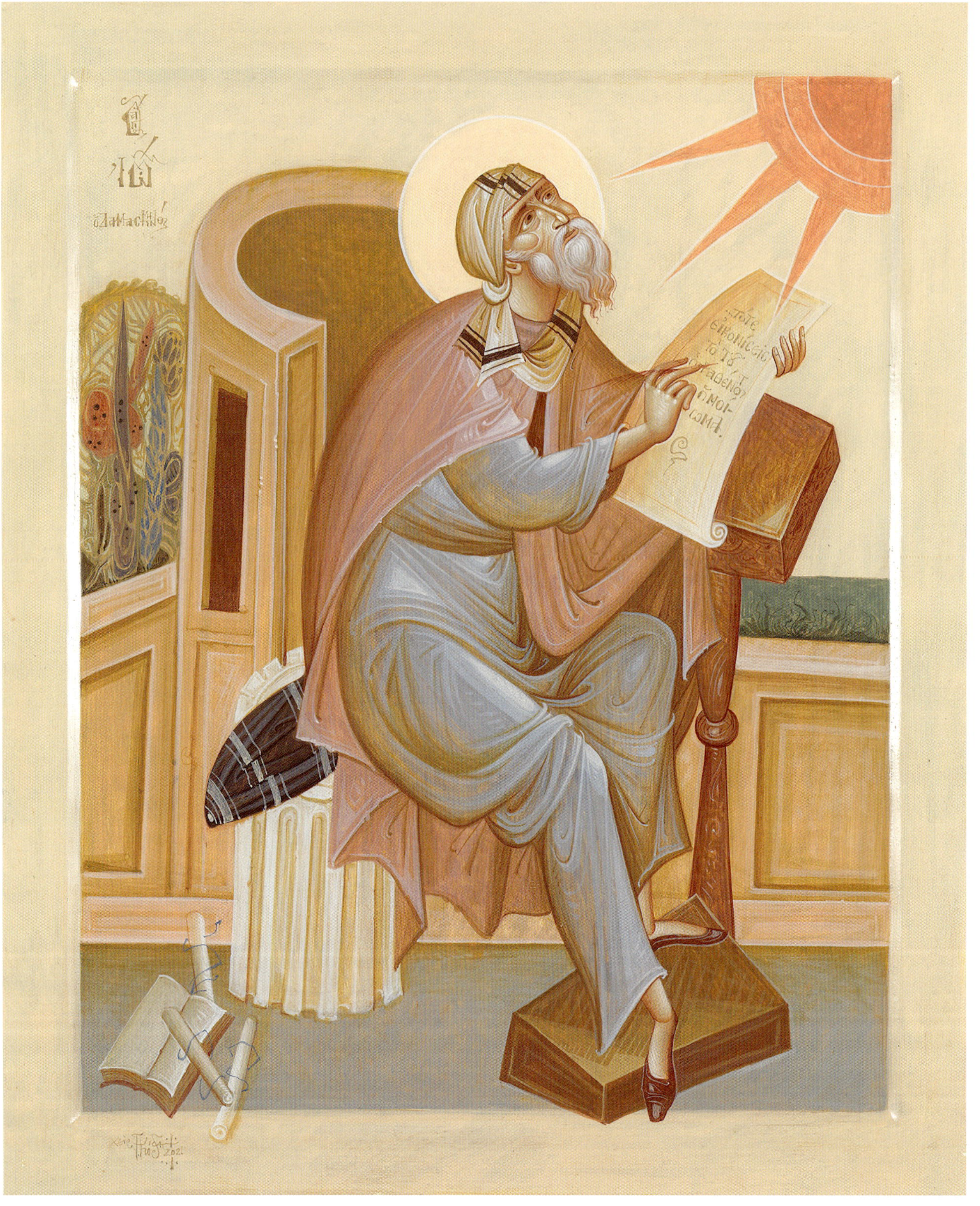

ST JOHN DAMASCENE,
egg tempera on wood, 50 x 60 cm

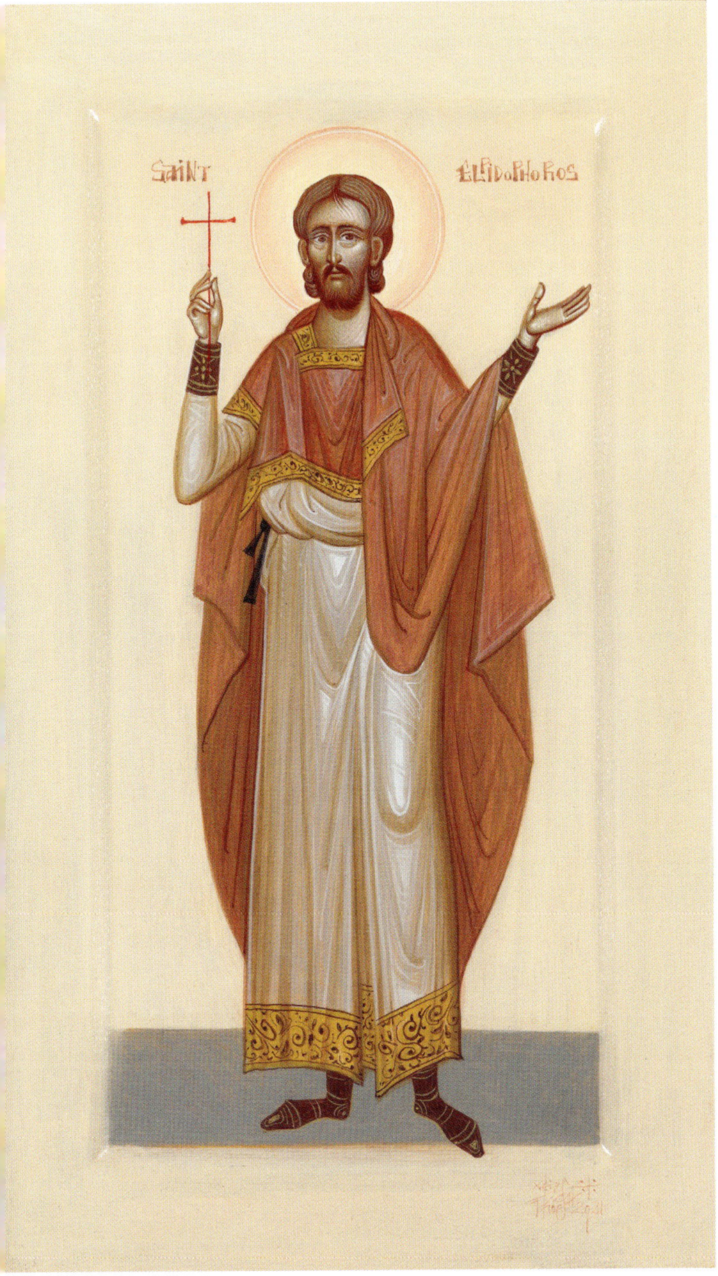

ST. ELPIDOFOROS,
egg tempera on wood, 25 x 54 cm

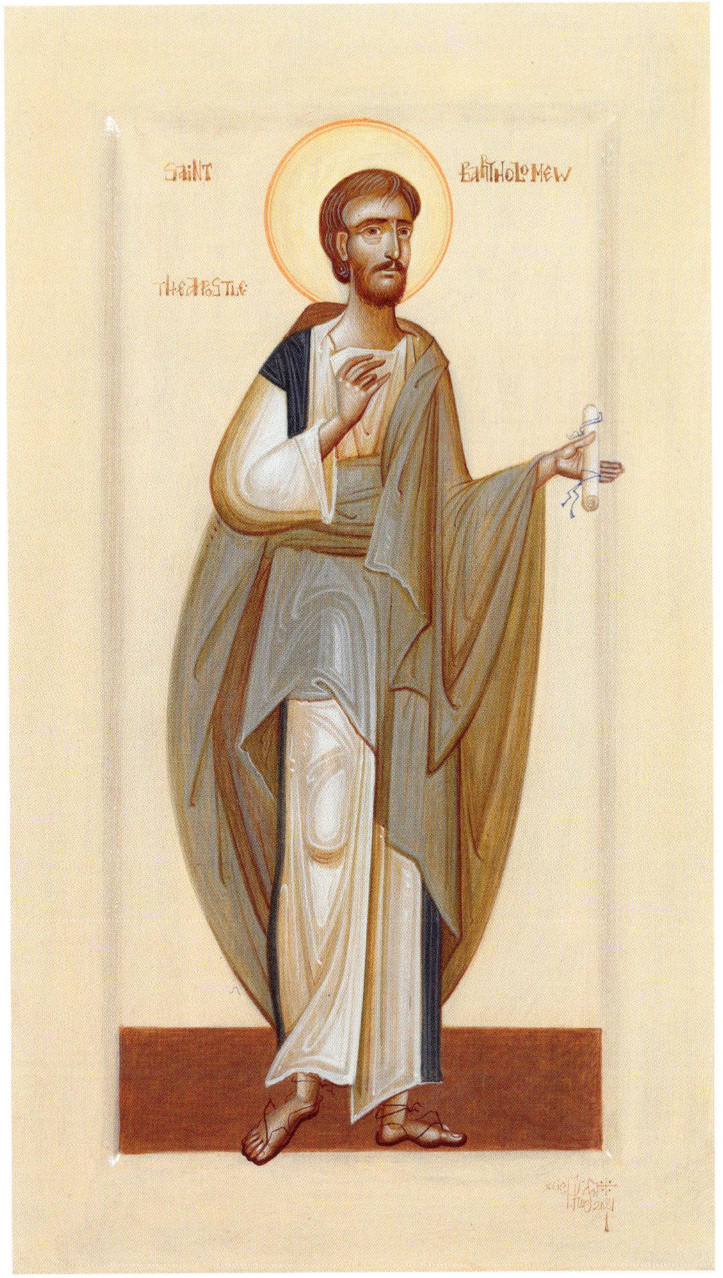

ST. BARTOLOMEOU THE APOSTLE,
egg tempera on wood, 25 x 54 cm

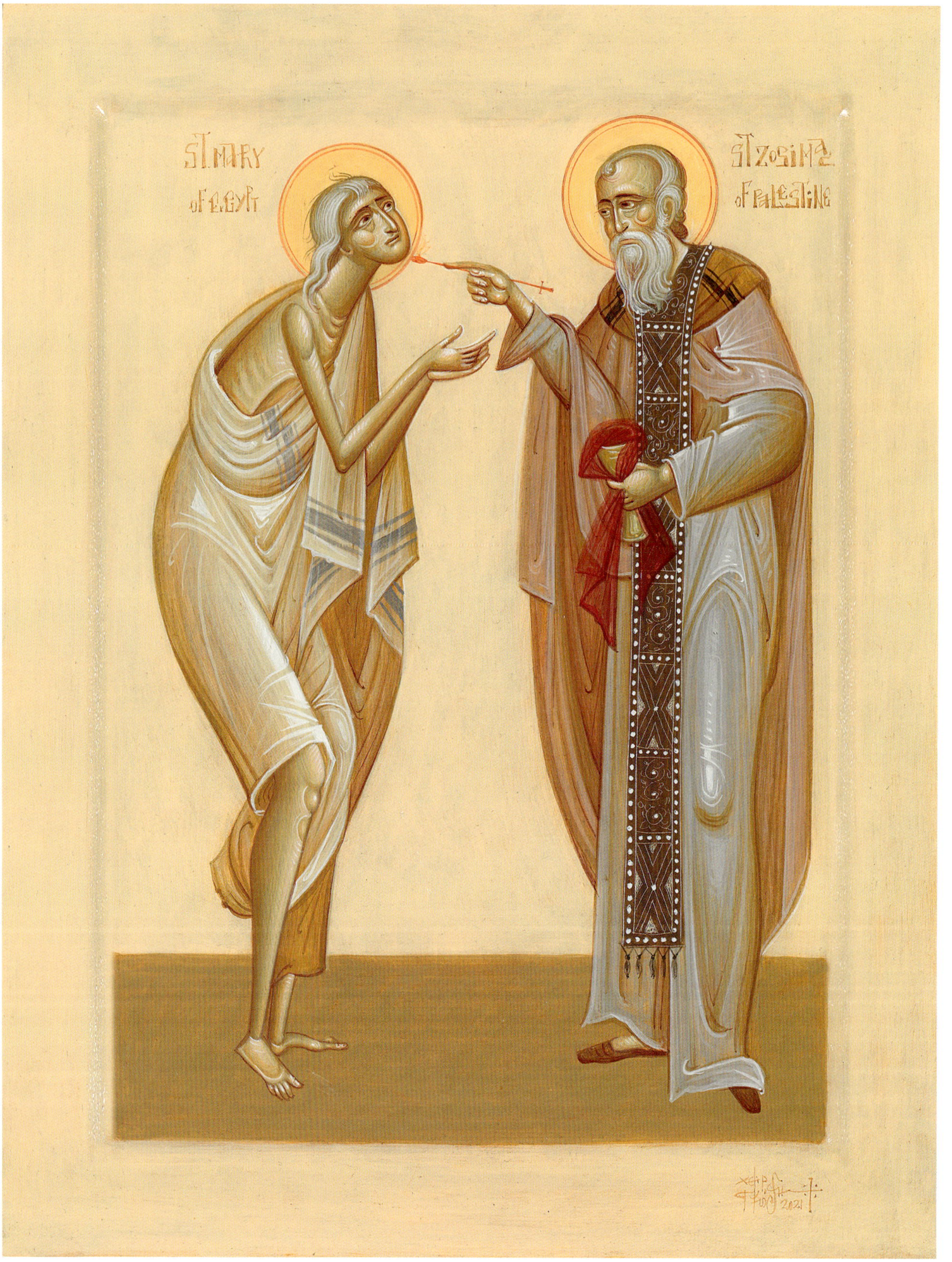

ST. MARY OF EGYPT,
egg tempera on wood, 40 x 50 cm

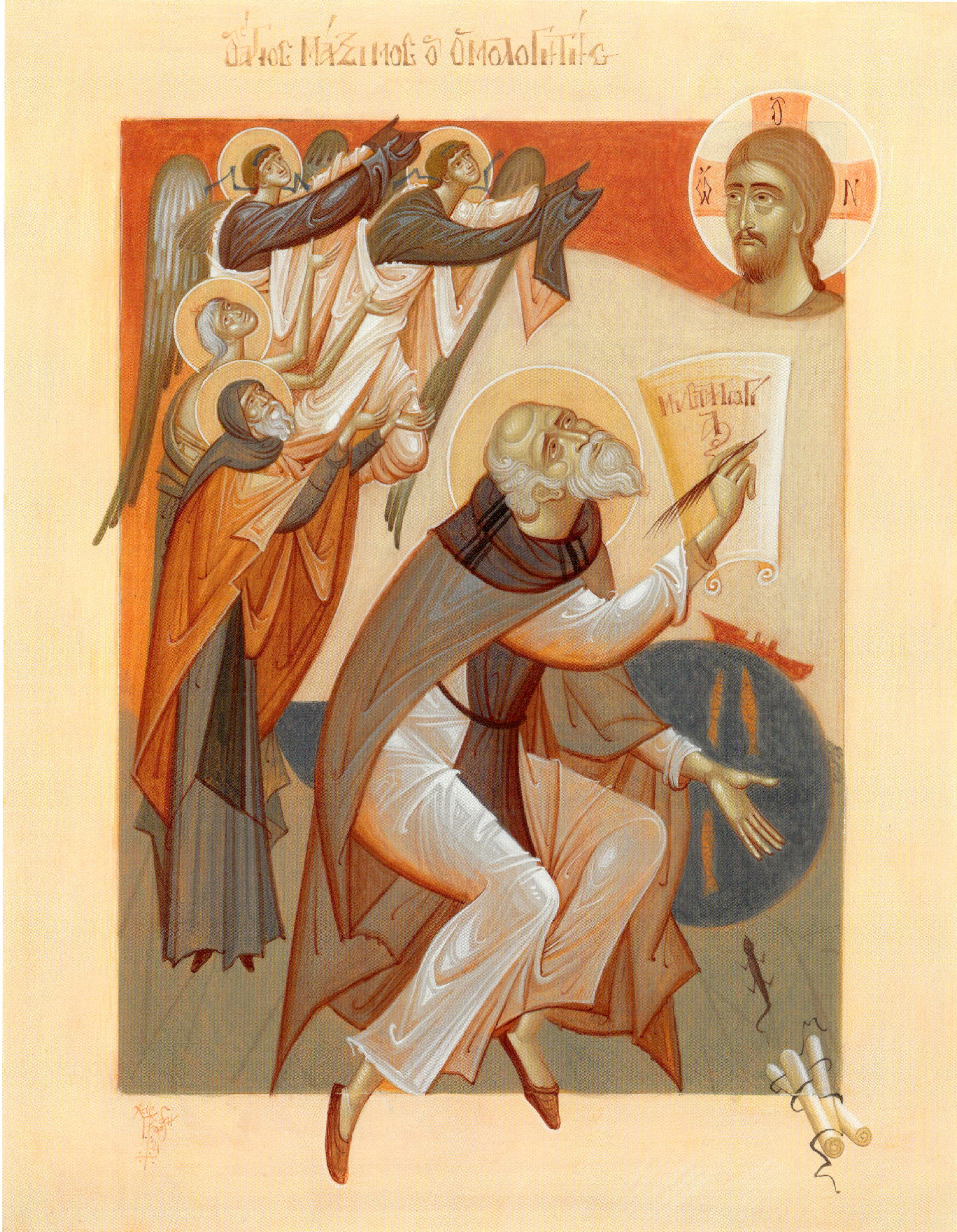

ST. MAXIMUS THE CONFESSOR,
egg tempera on board, 40 x 50 cm

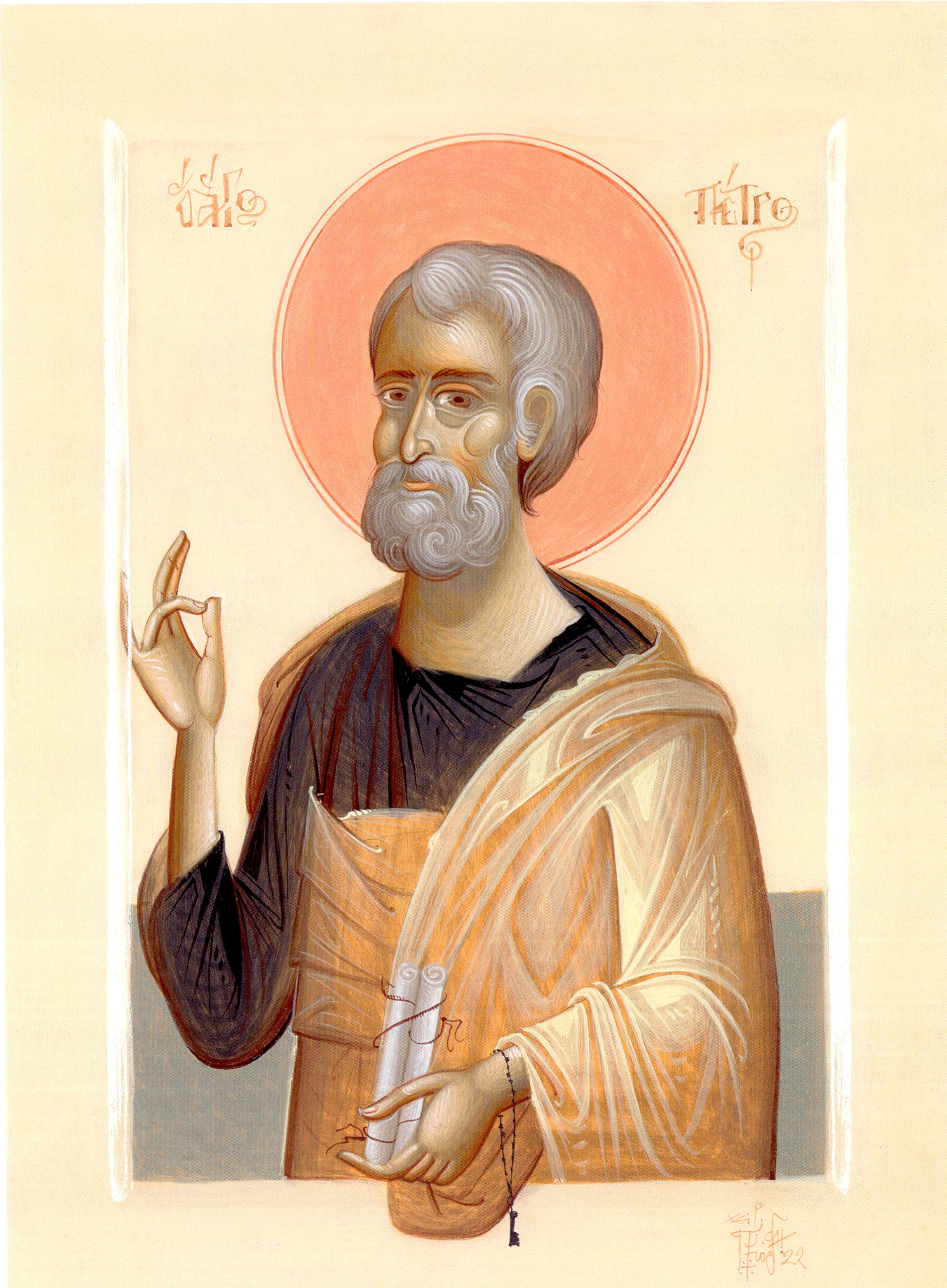

ST. PETER, egg tempera on board, 40 x 30 cm

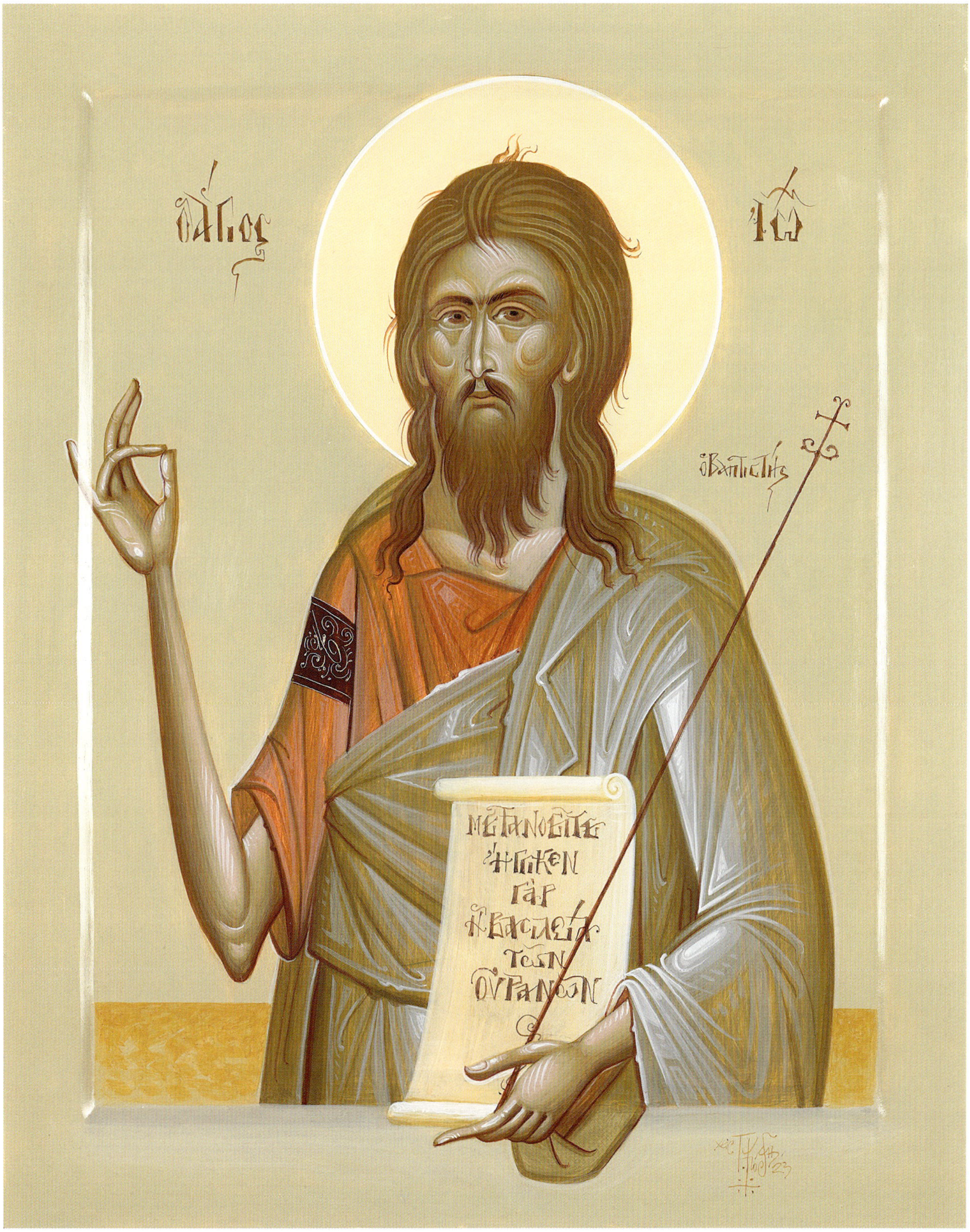

ST. JOHN THE BAPTIST, egg tempera on board, 45 x 55 cm

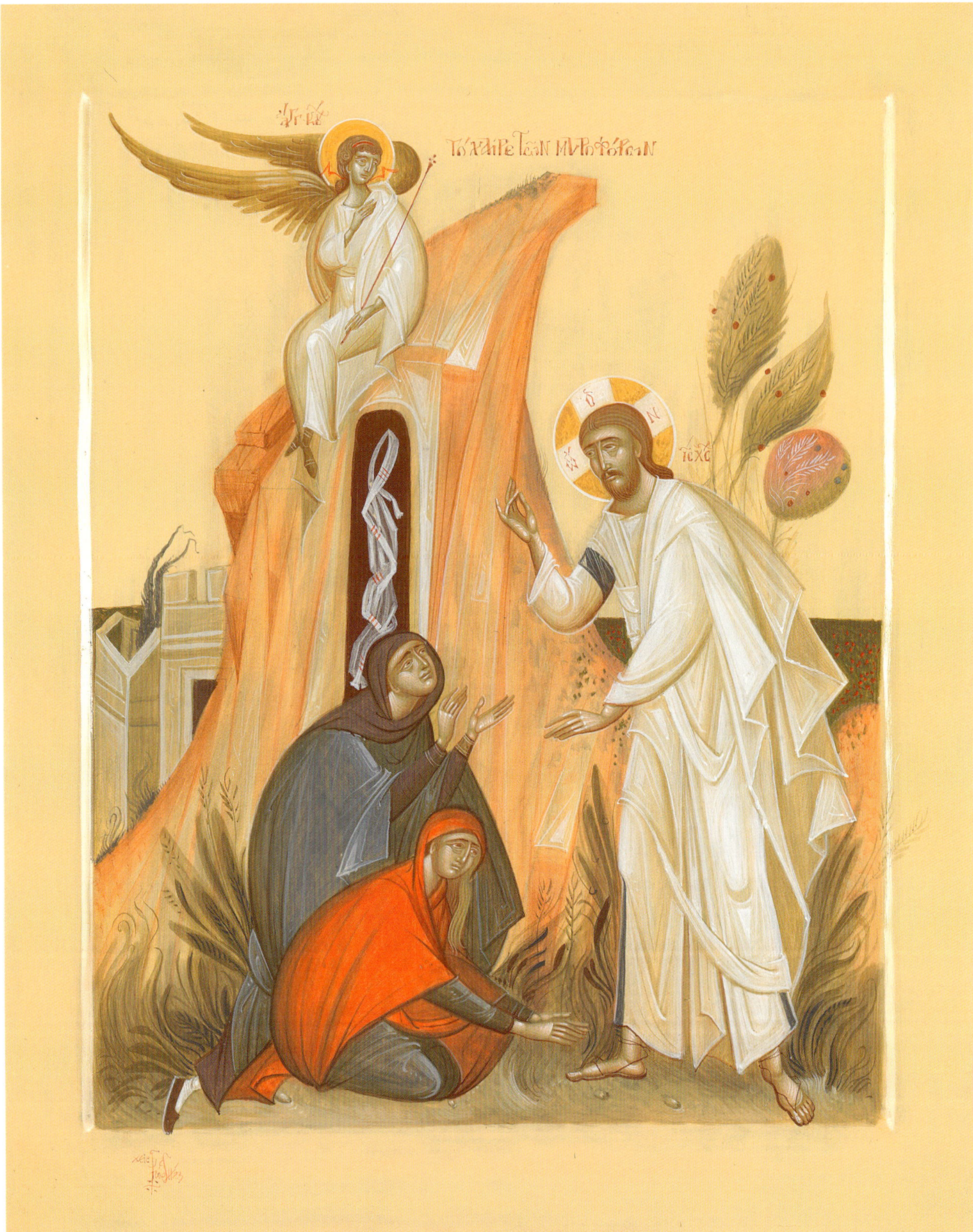

THE MYHRR BEARER WOMEN,
egg tempera on board, 45 x 55 cm

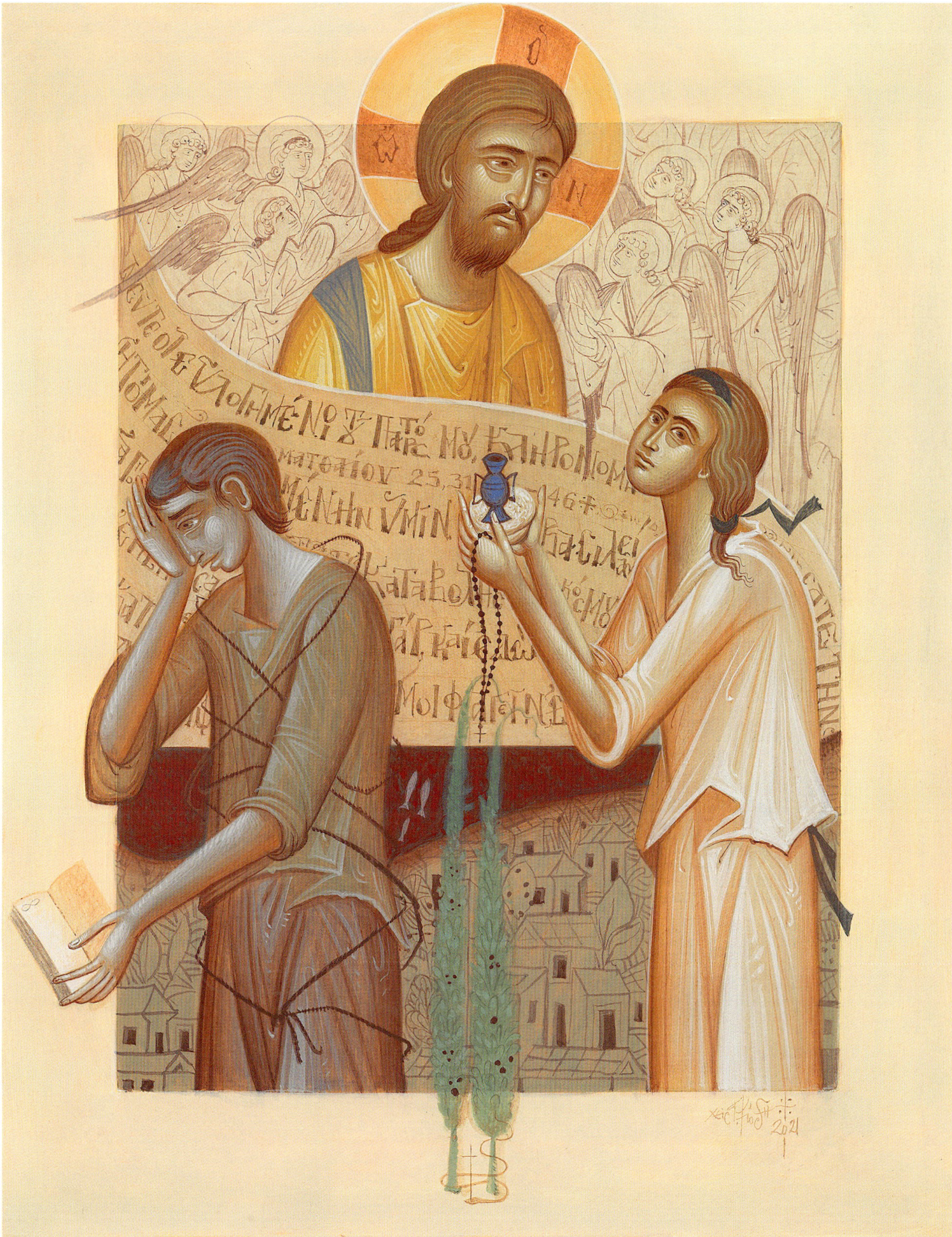

THE PARABLE OF LAST JUDGMENT,
egg tempera on board, 40 x 50 cm

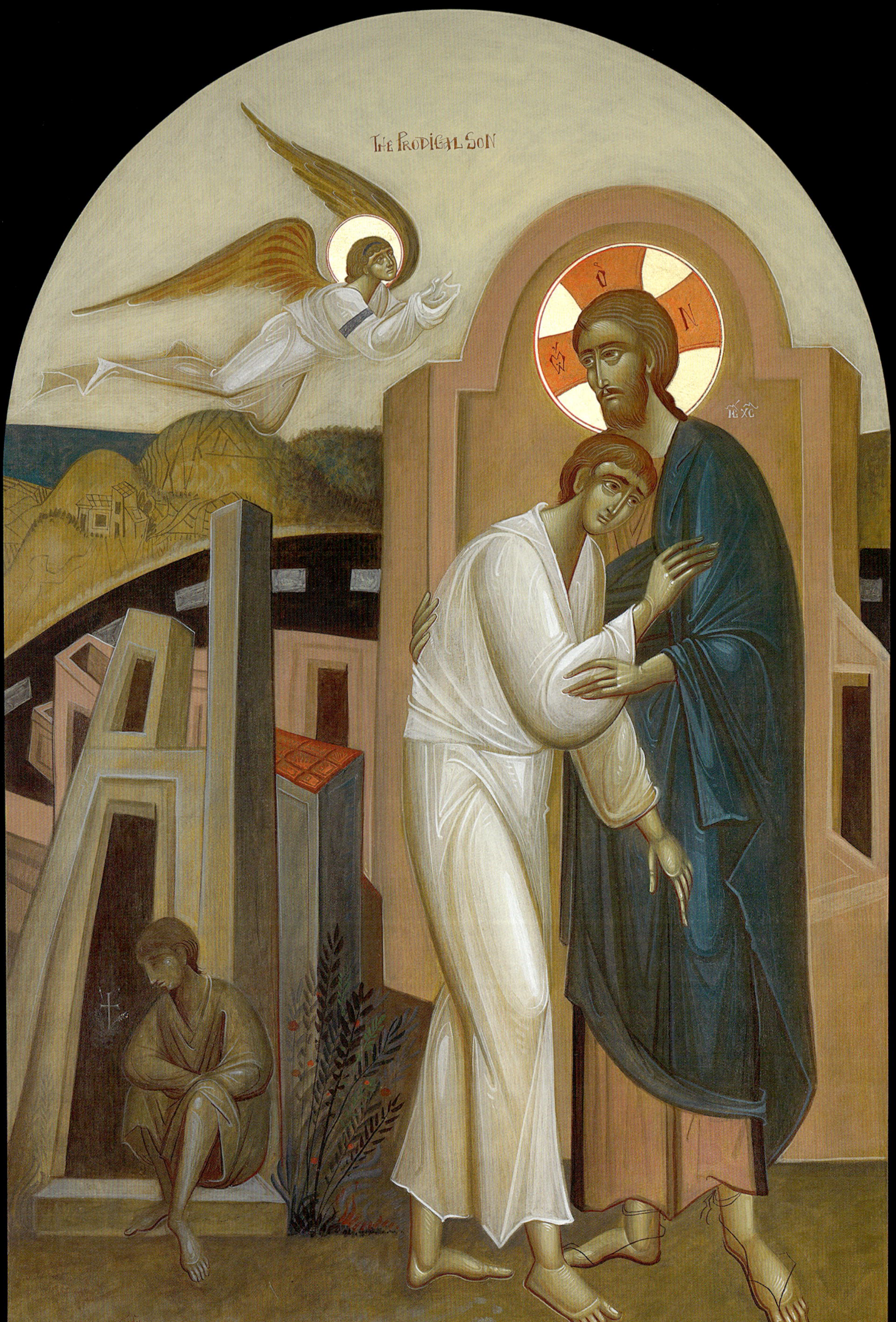
THE PRODIGAL SON

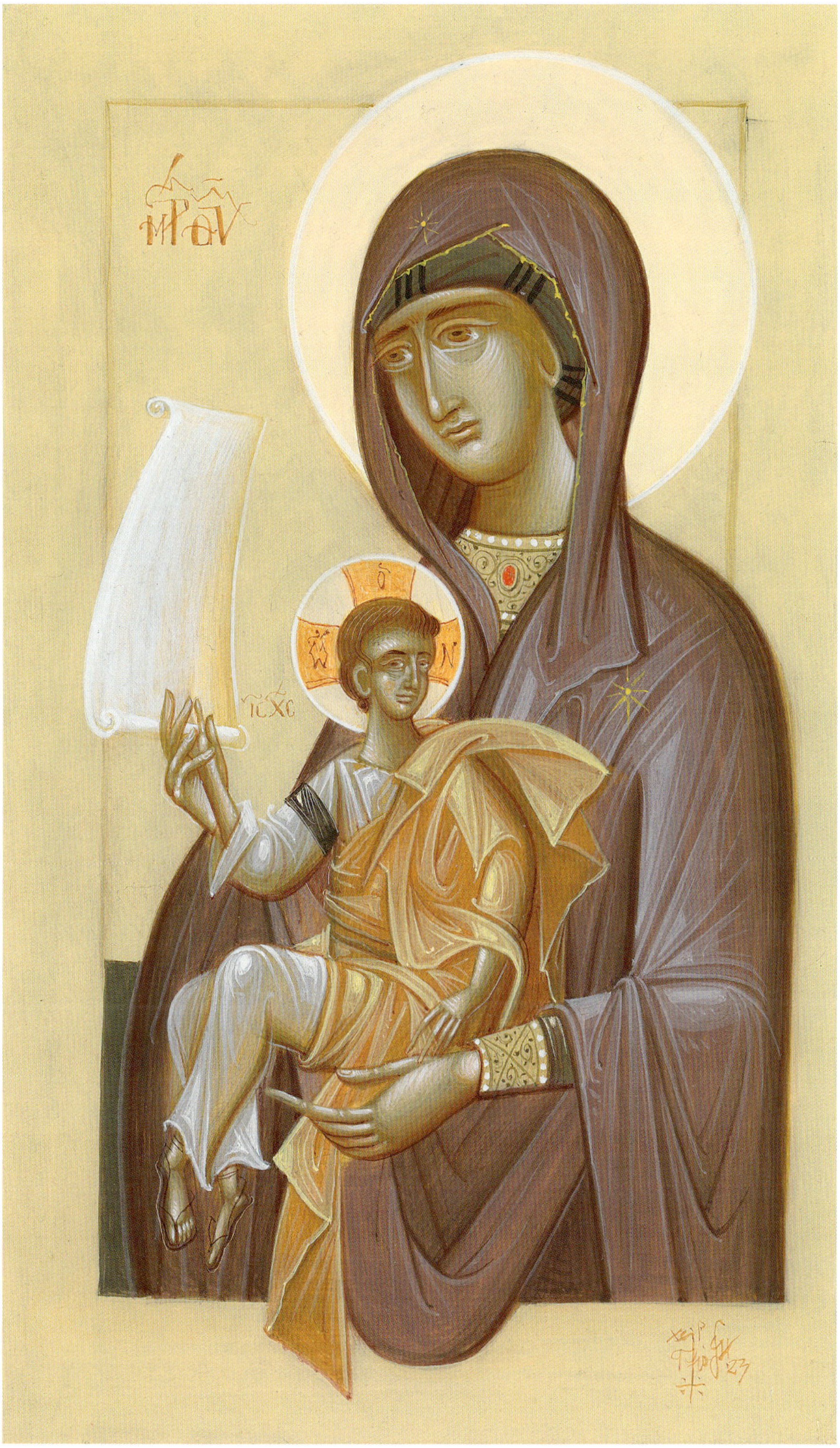

THEOTOKOS THE AXION ESTIN,
egg tempera on board,
30 x 40 cm

E PRODIGAL SON,
xed media on board,
x 150 cm

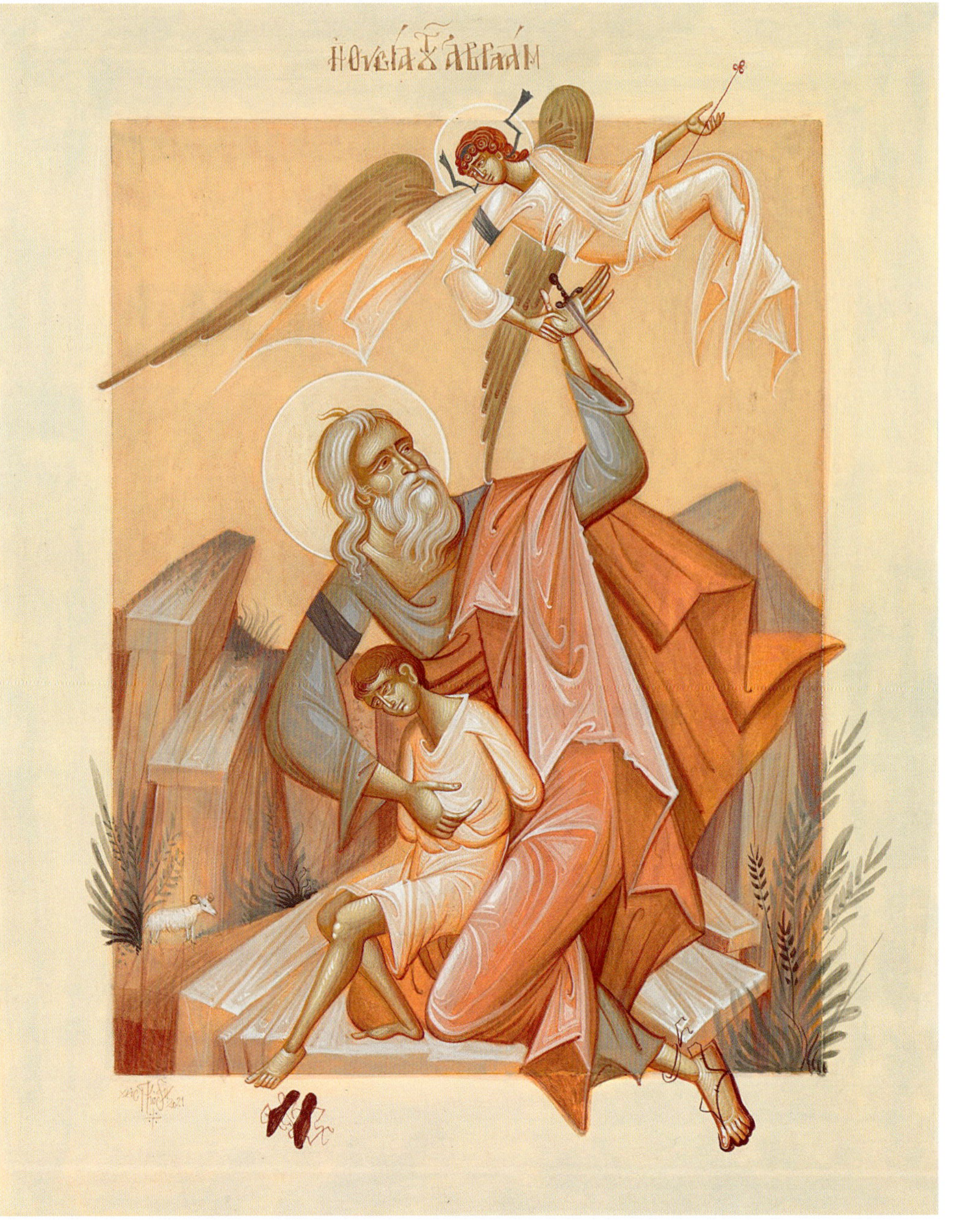

THE SACRIFICE OF ABRAHAM,
egg tempera on board,
40 x 50 cm

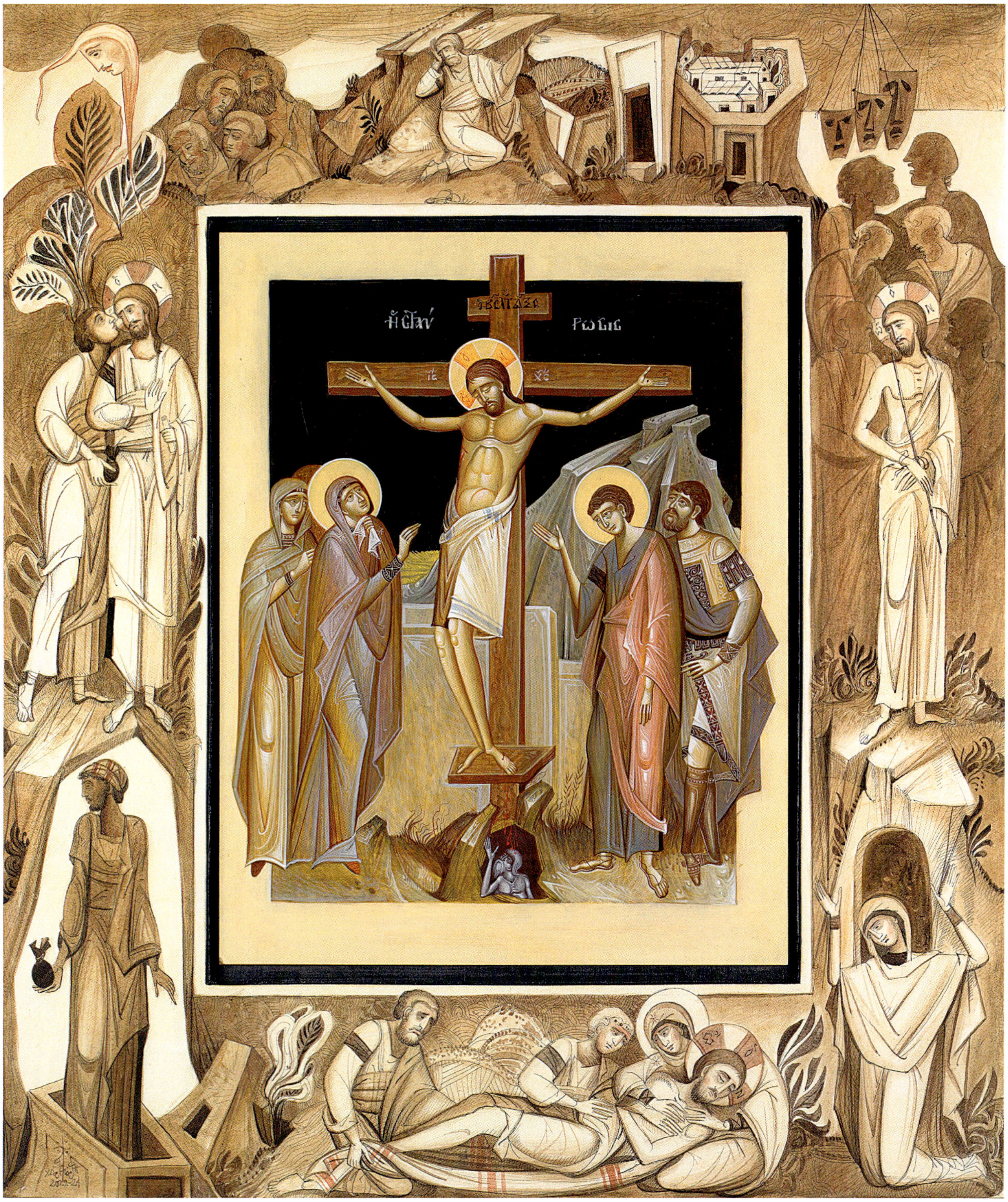

CRUCIFIXION, egg tempera and inks of wood, 60 x 90 cm

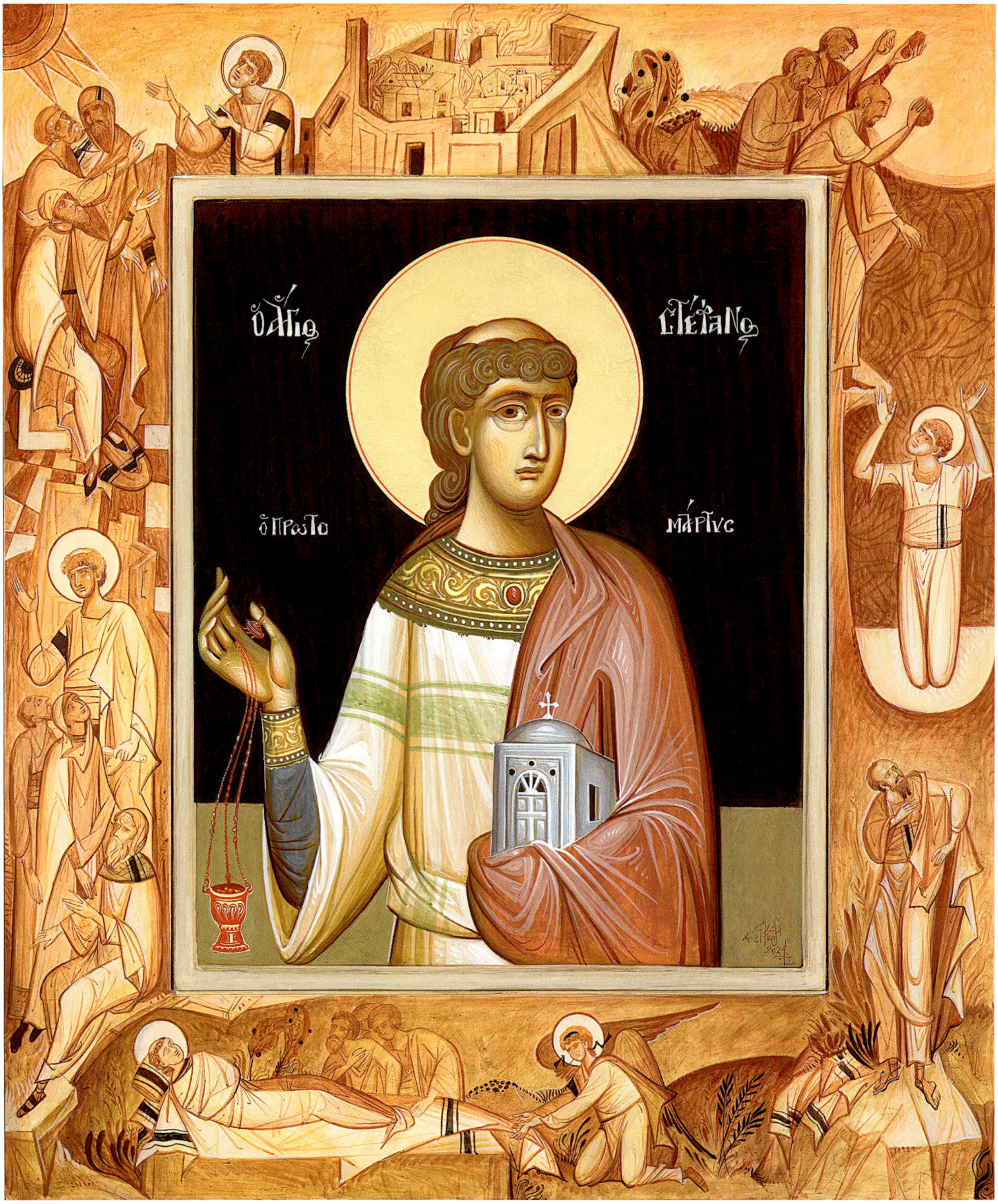

ST. STEPHEN WITH SCENES OF HIS LIFE,
mixed media on wood, 50 x 60 cm

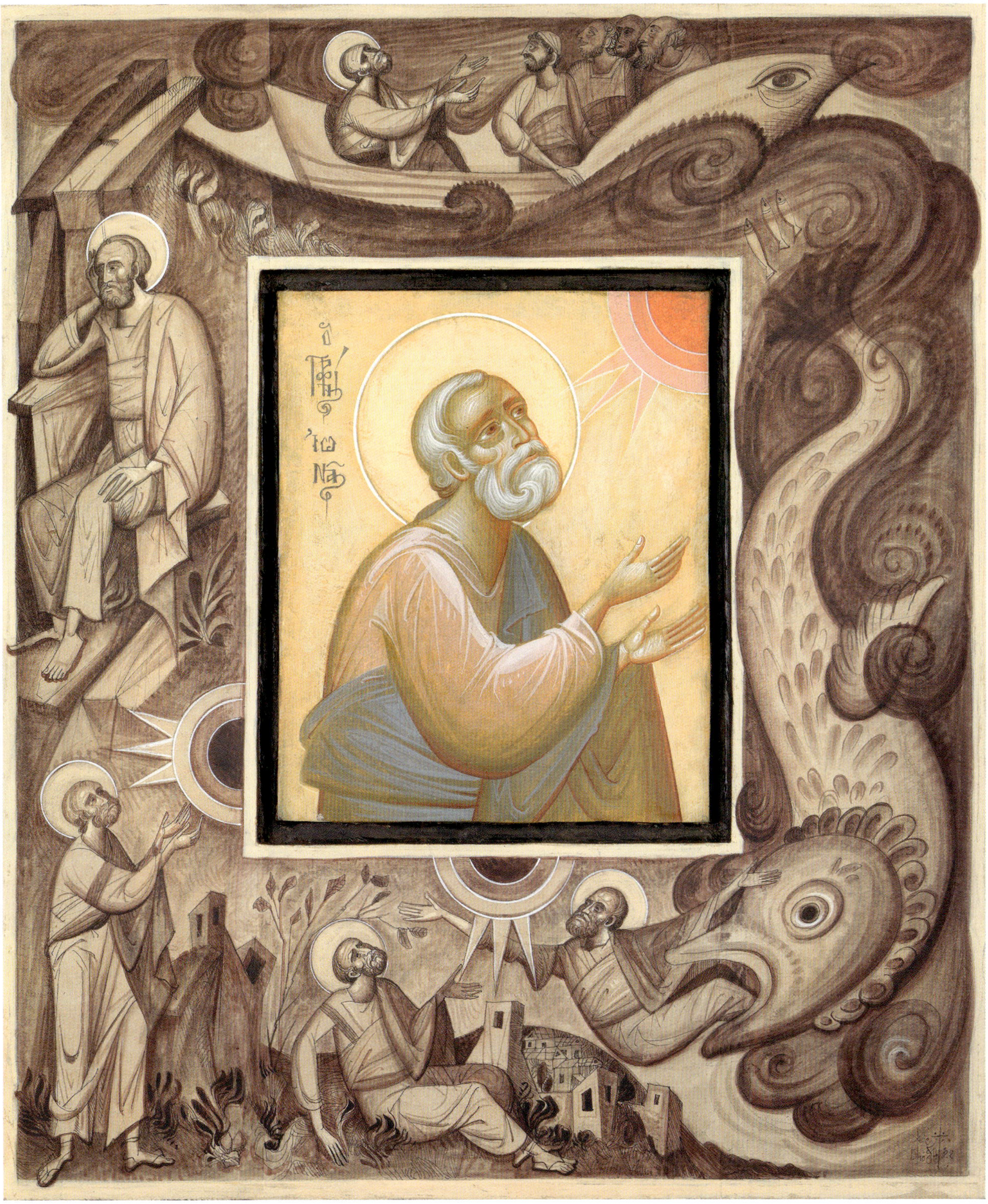

PROPHET JONAS WITH SCENES OF HIS LIFE,
egg tempera and inks on wood, 50 x 60 cm

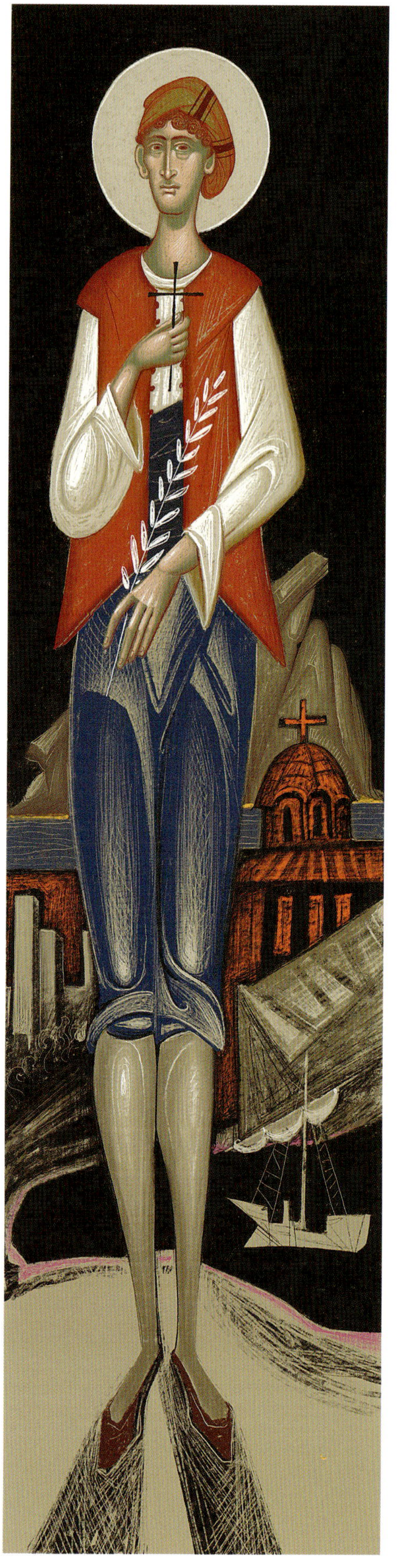

SAINT TRIANTAFYLLOS,
digital painting,
50 x 160 cm

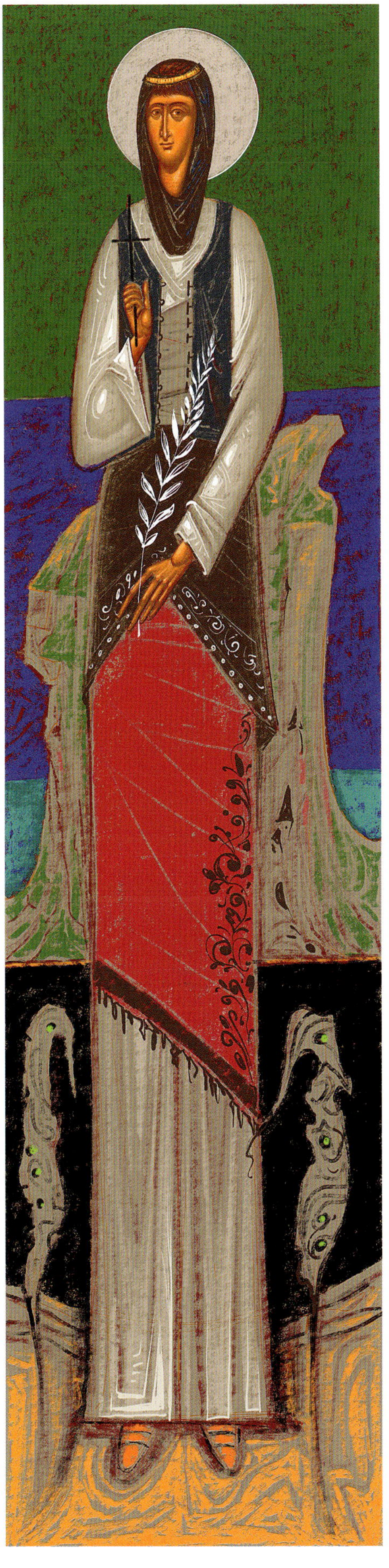

ST HELEN OF SINOPE,
digital painting,
50 x 160 cm

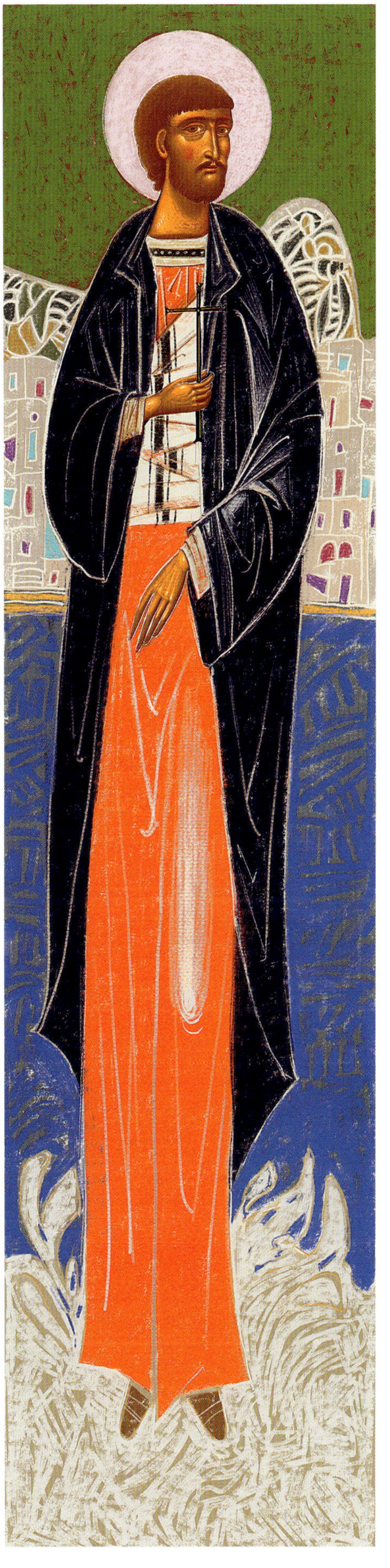

ST. CONSTANTINE
OF HYDRA,
digital painting,
50 x 160 cm

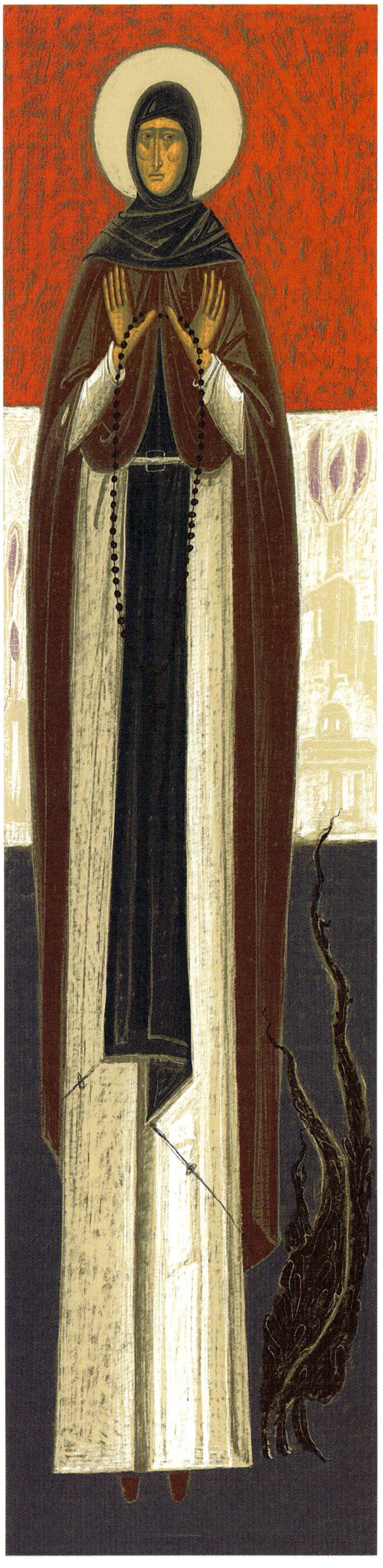

ST. FILOTHEA, digital painting, 50 x 160 cm

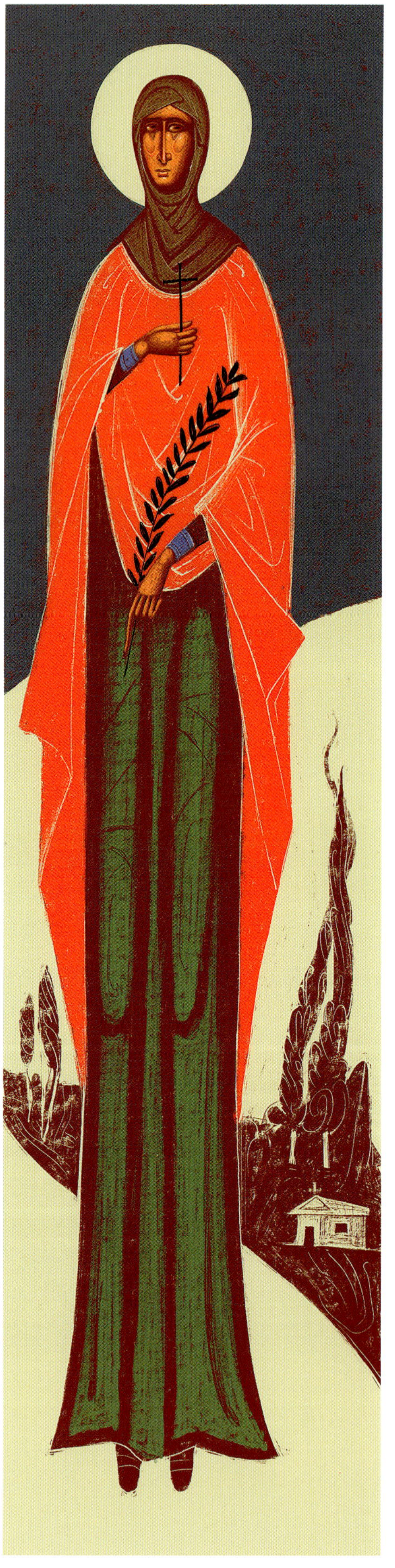

ST. KYRANNA, digital painting, 50 x 160 cm

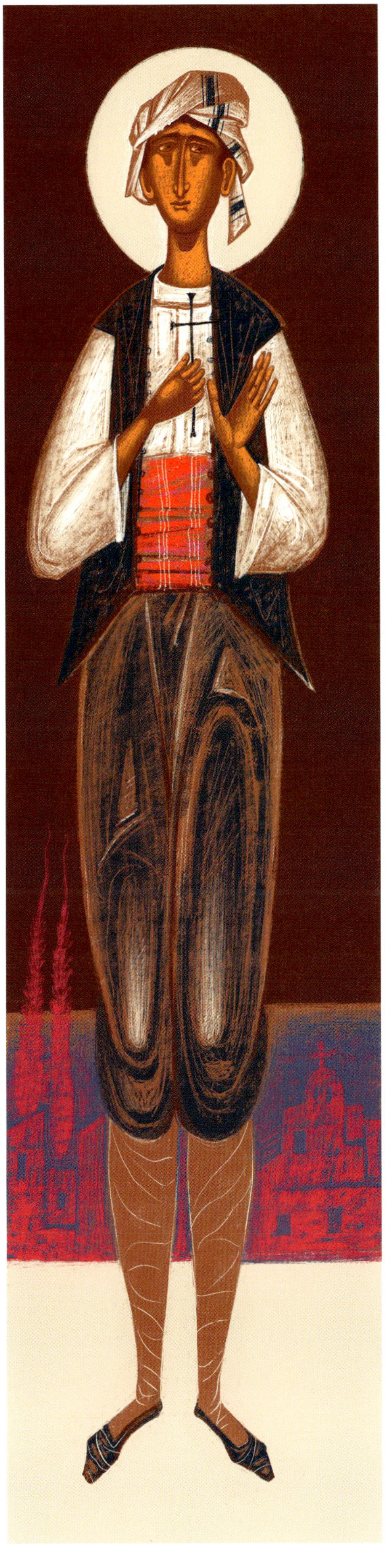

ST. NIKITAS FROM NISYROS ISLAND, *digital painting, 50 x 160 cm*

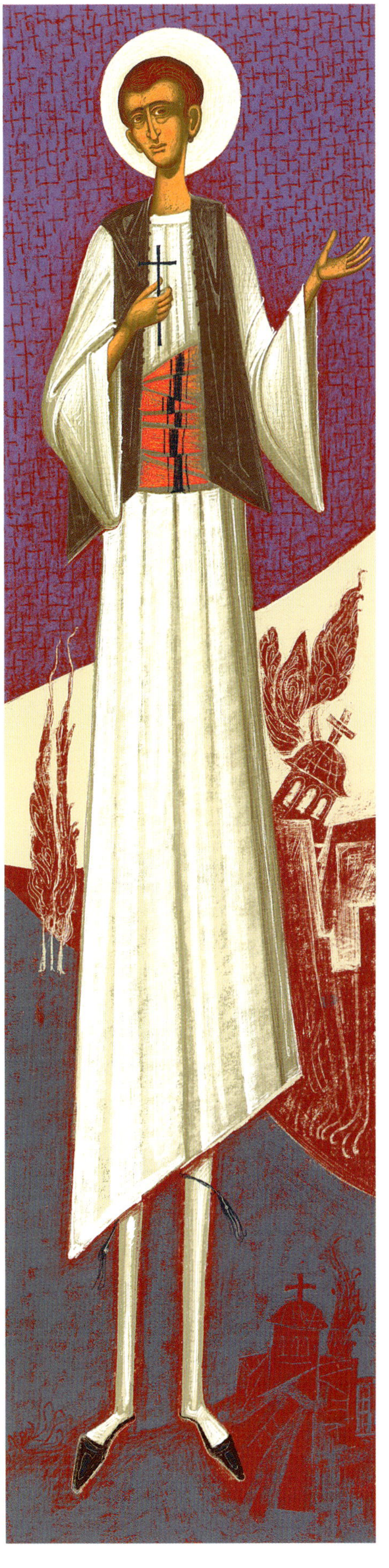

SAINT JOHN FROM MONEMVASIA, *digital painting, 50 x 160 cm*

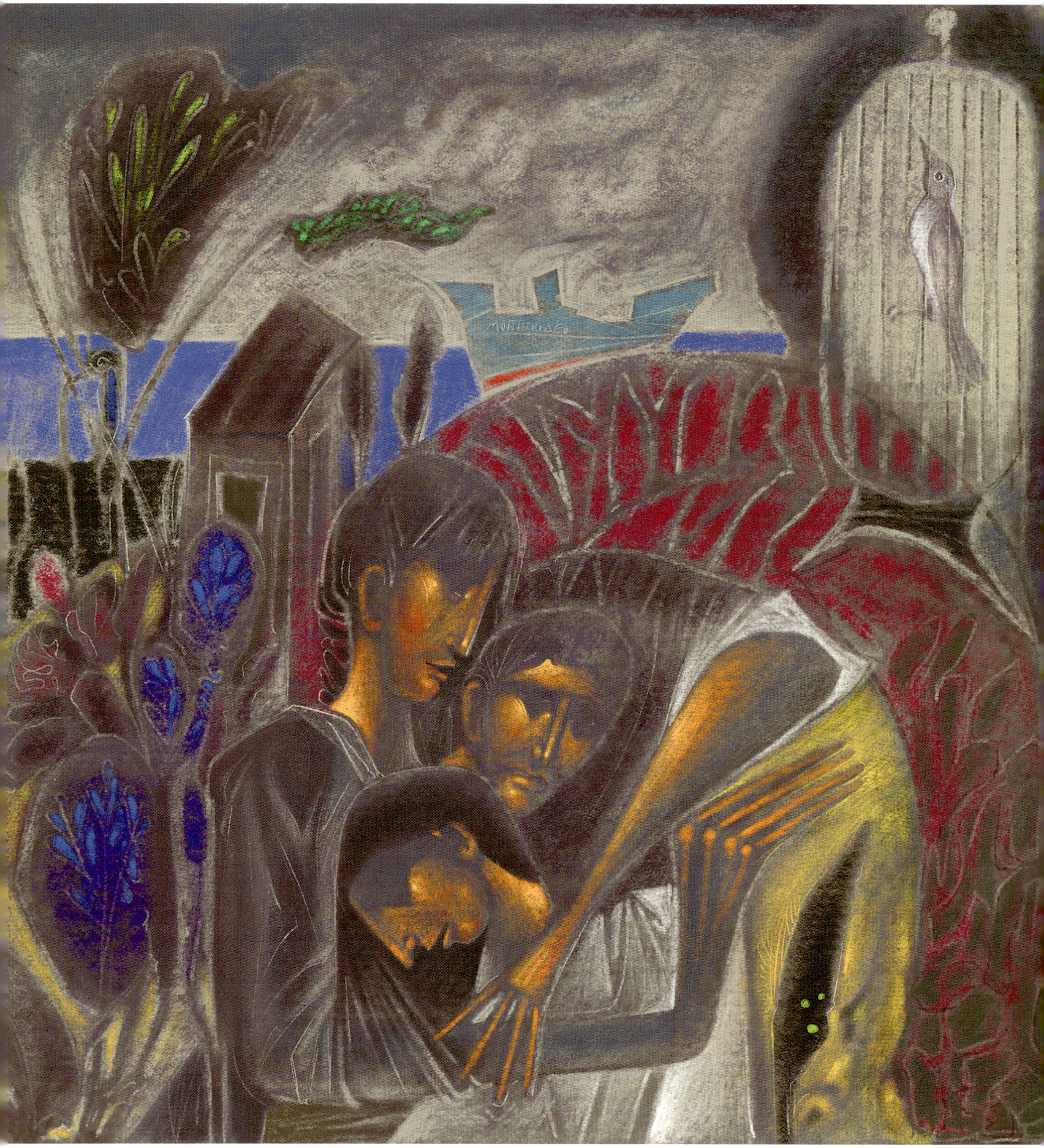

A SHIP CALLED MONTEVIDEO,
digital painting, 120 x 120 cm

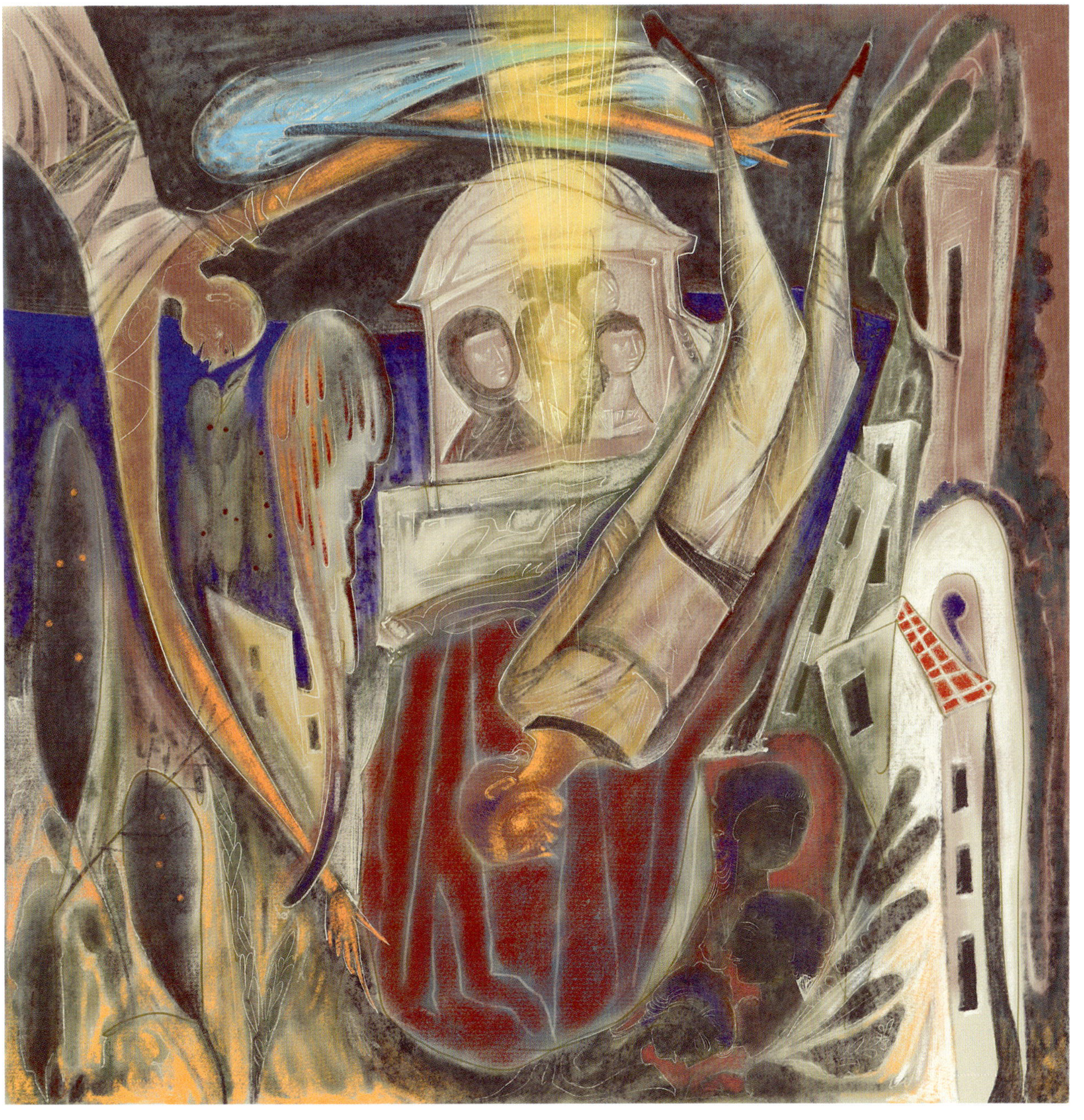

PEOPLE FLY IN MONTEVIDEO,
digital painting, 120 x 120 cm

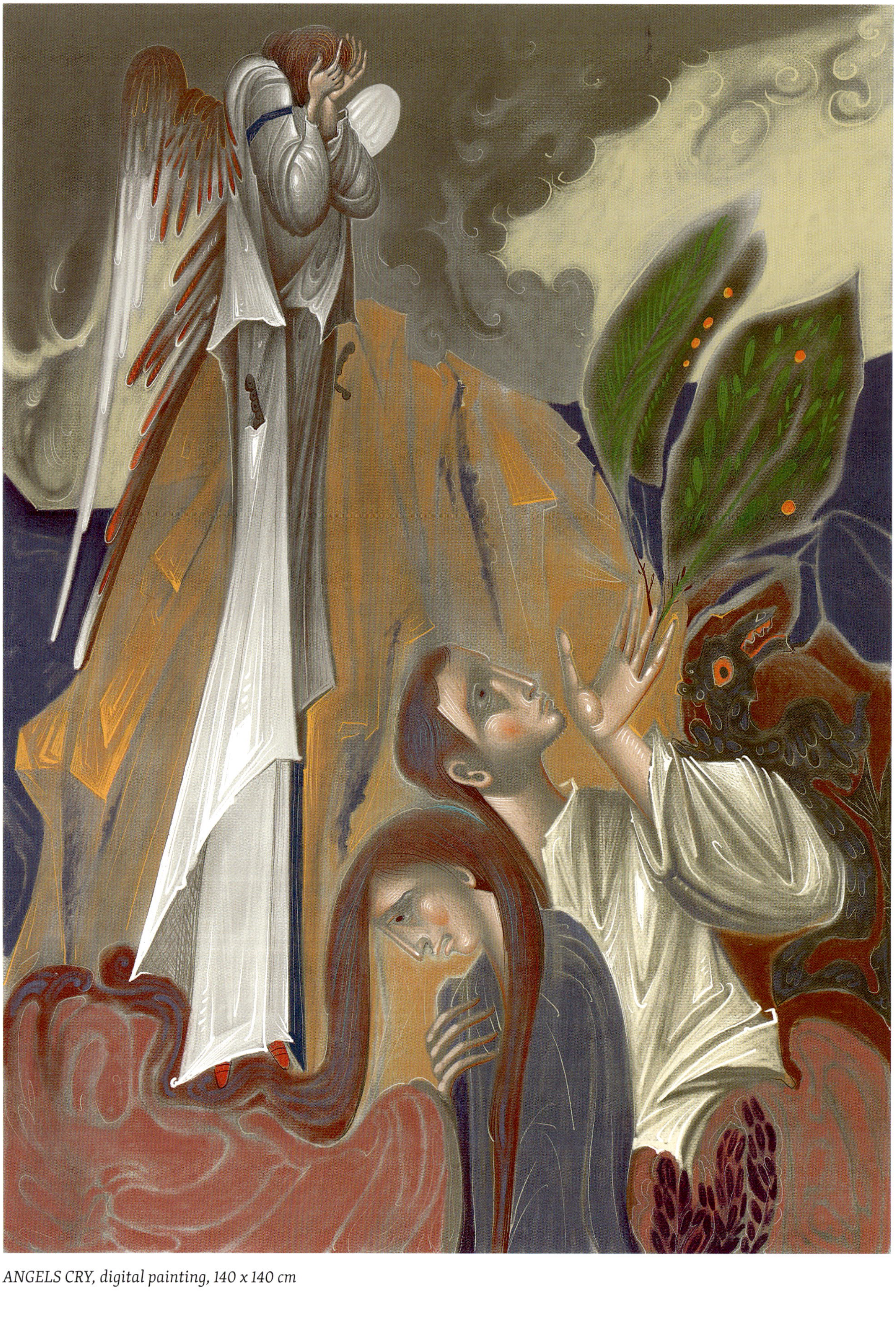

ANGELS CRY, digital painting, 140 x 140 cm

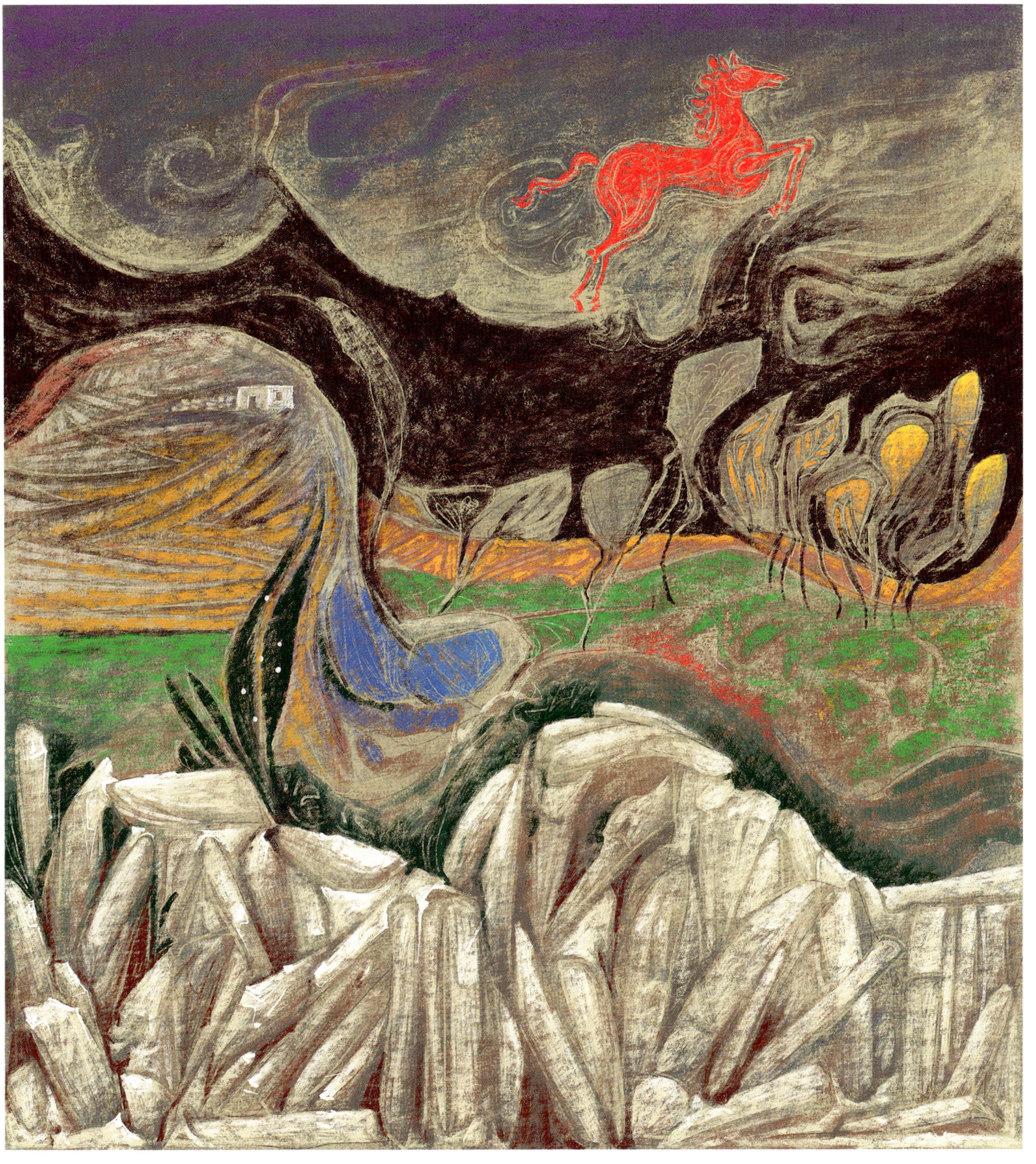

FARYTALE IN IRELAND, digital painting, 140 x 140 cm

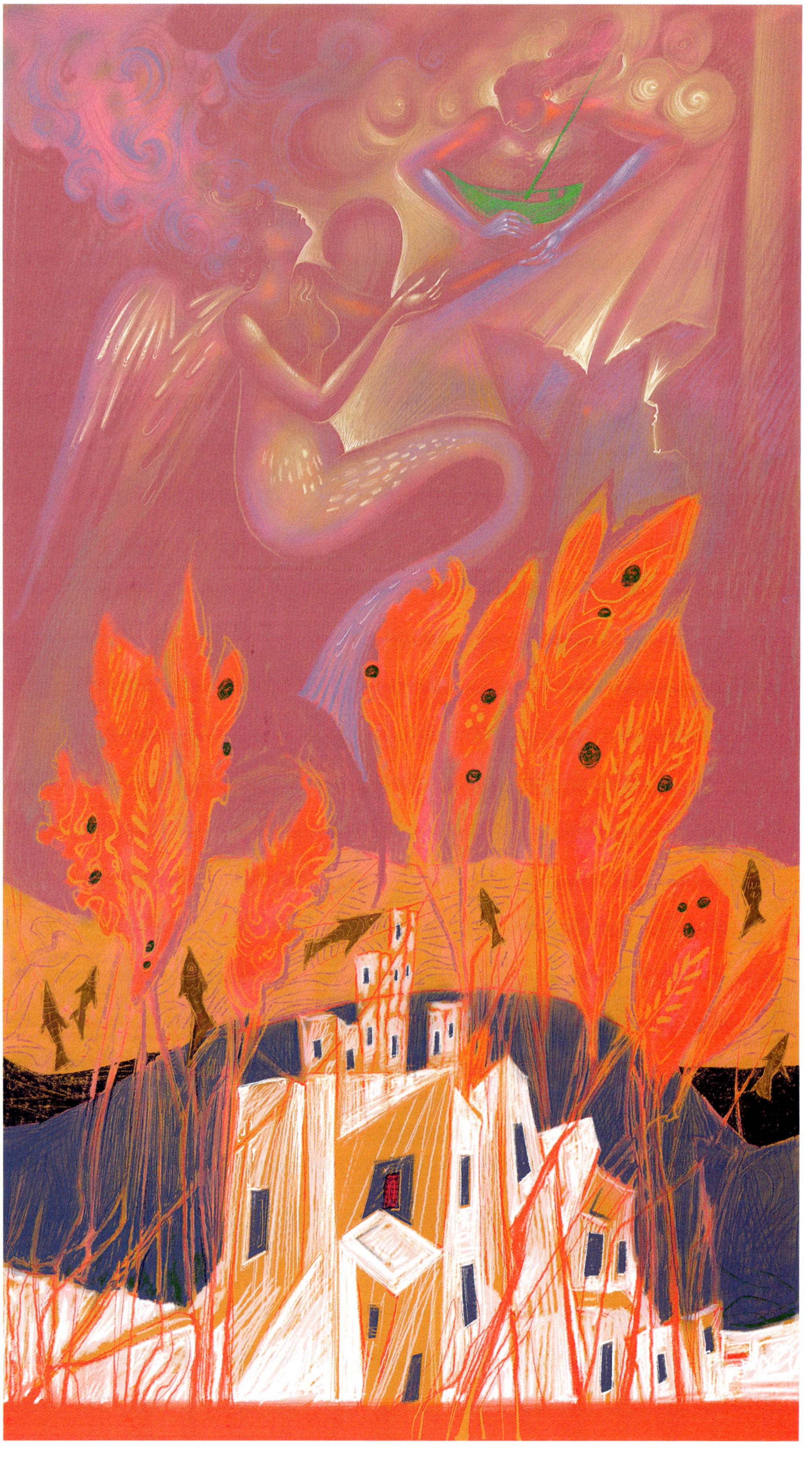

GREEN BOAT,
digital painting,
120 x 180 cm

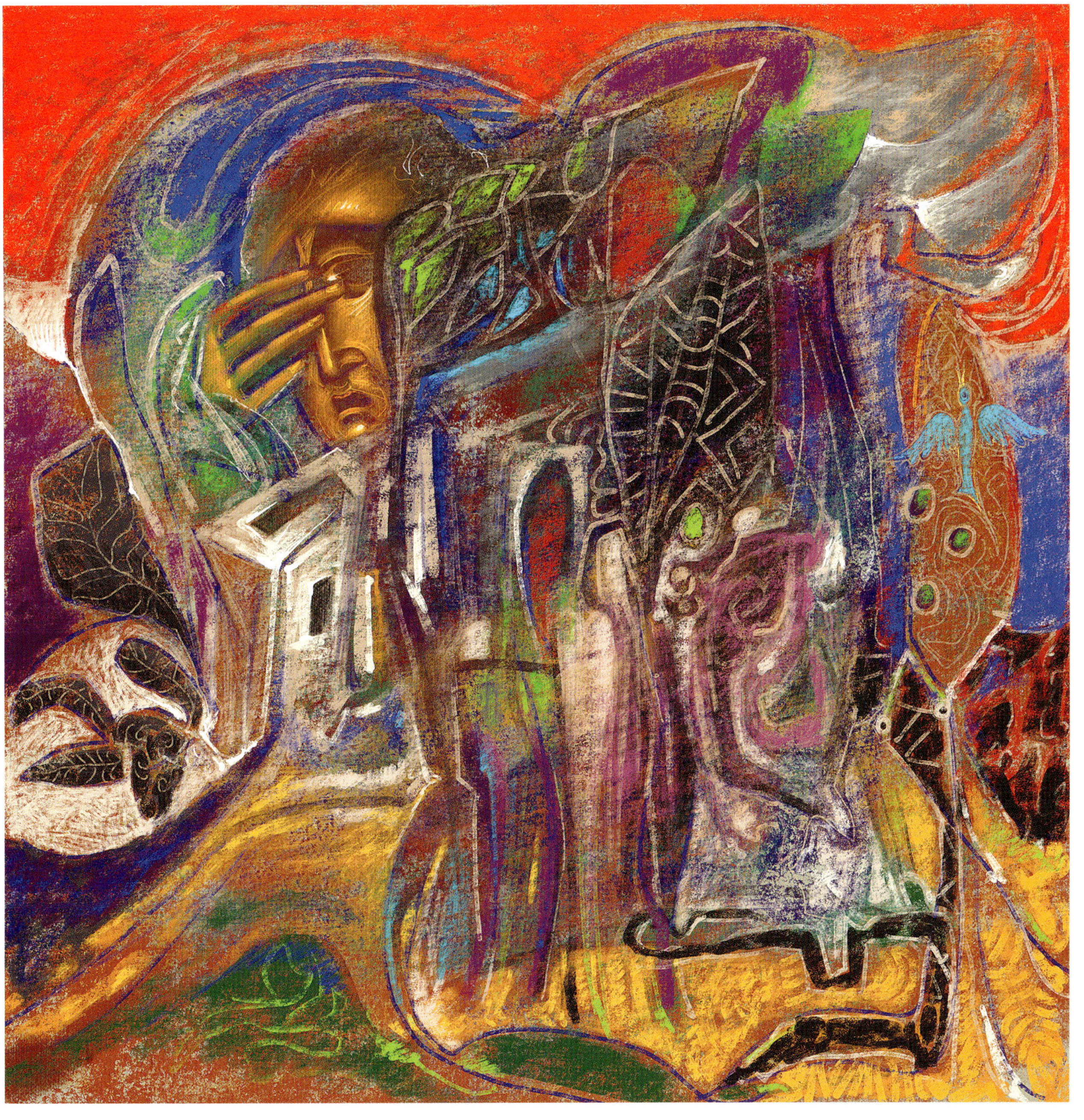

HIDDEN LIFE, digital painting, 140 x 140 cm

LIBATION,
digital painting,
120 x 140 cm

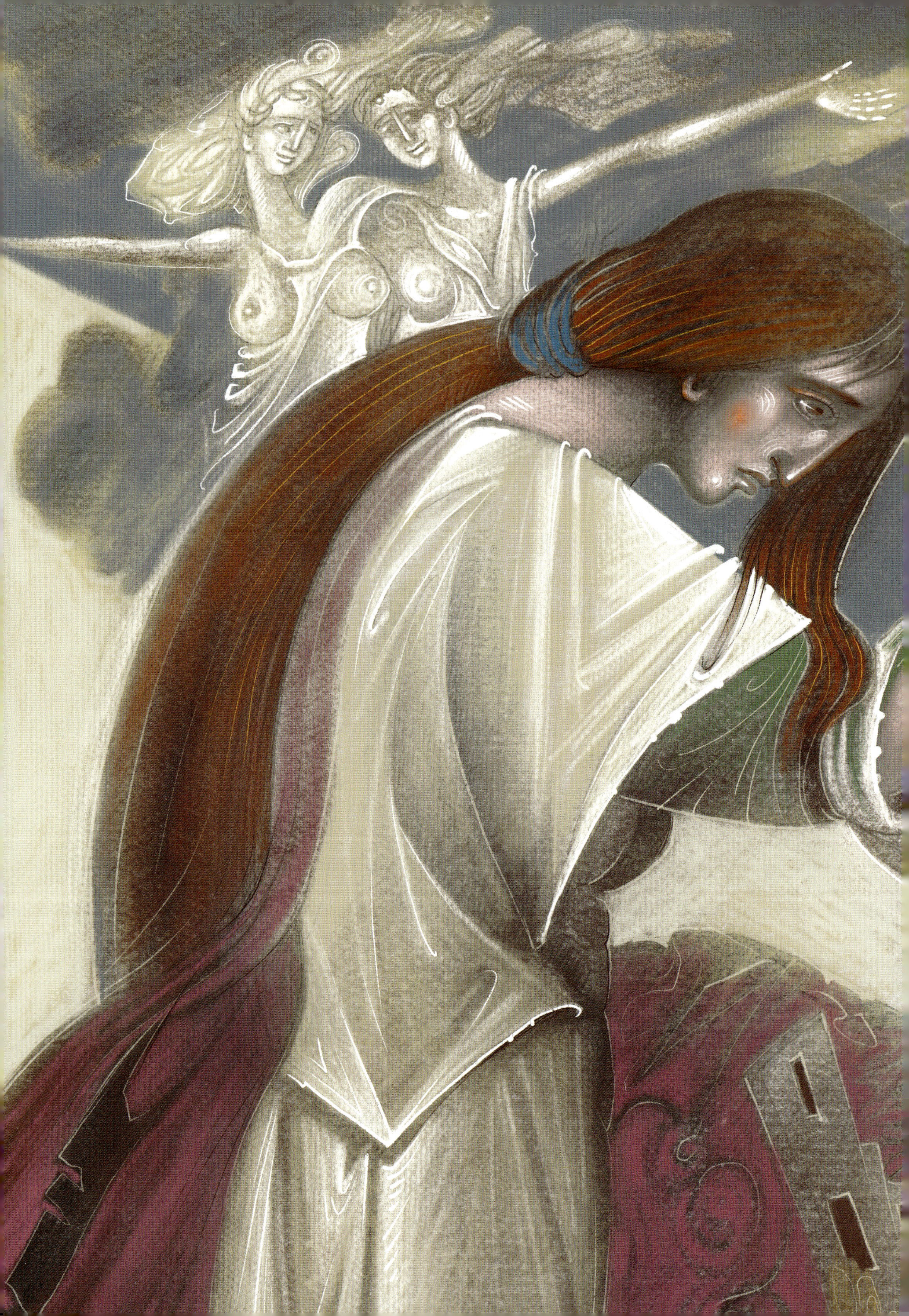

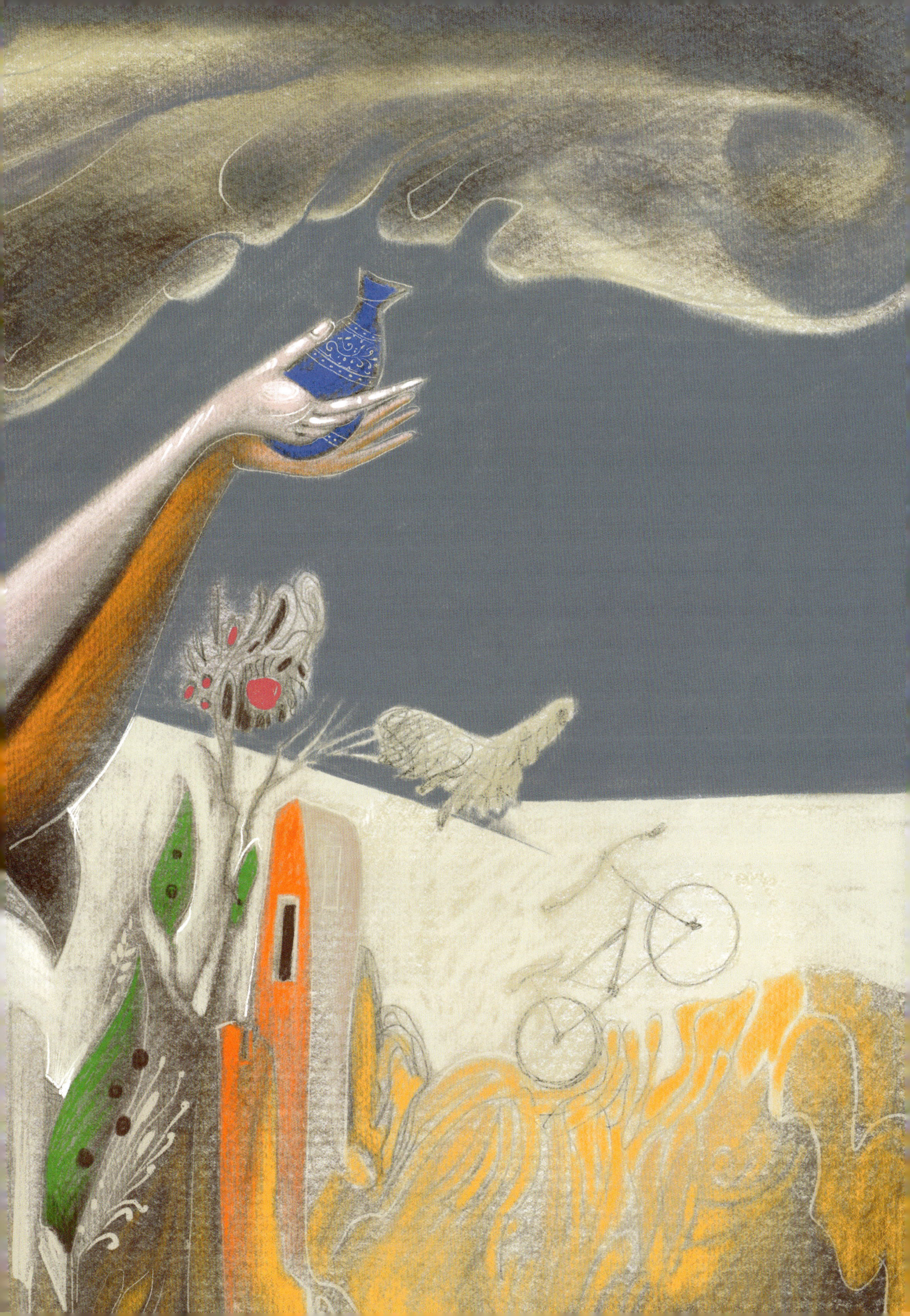

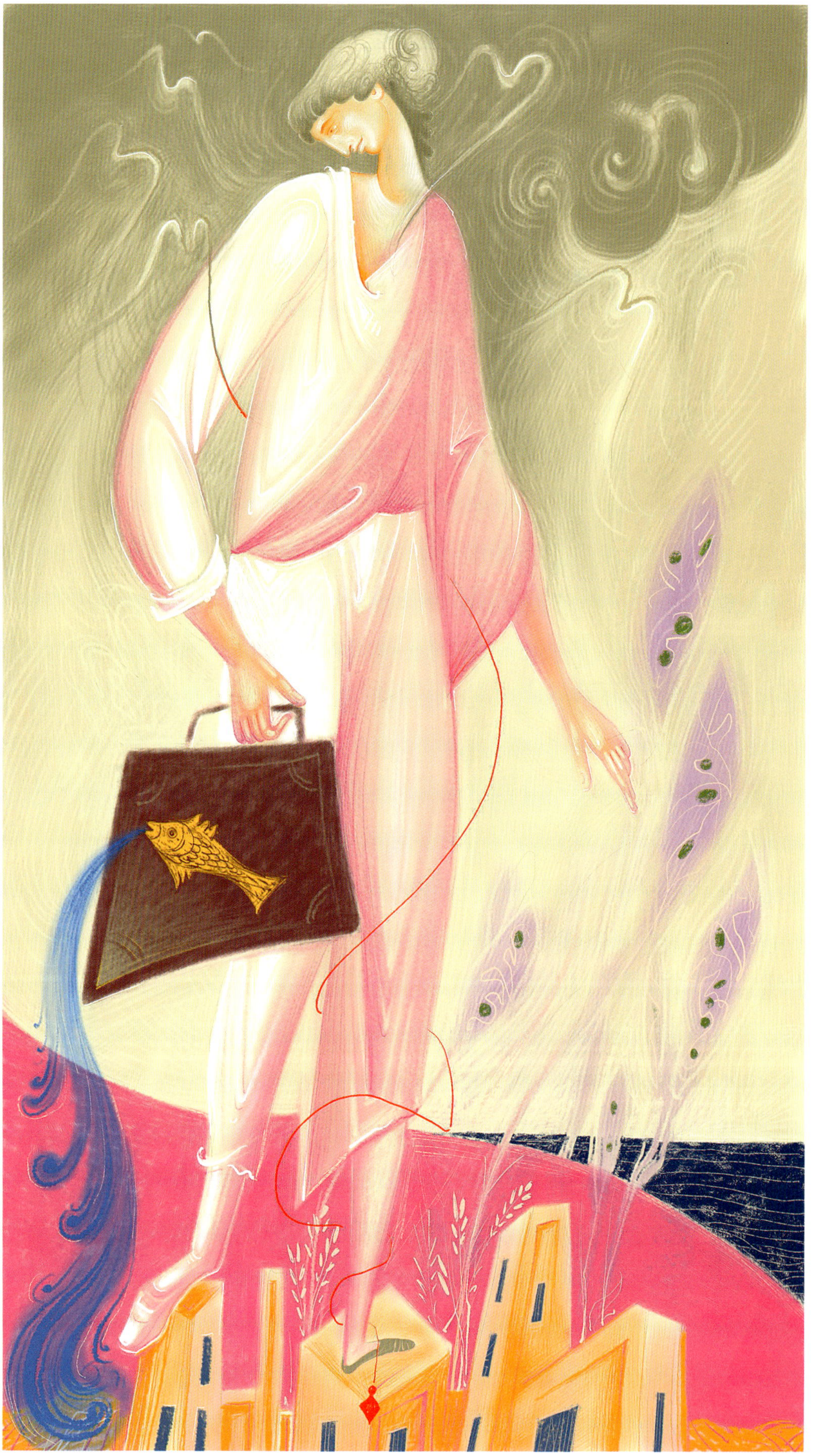

PILGRIM,
digital painting,
120 x 180 cm

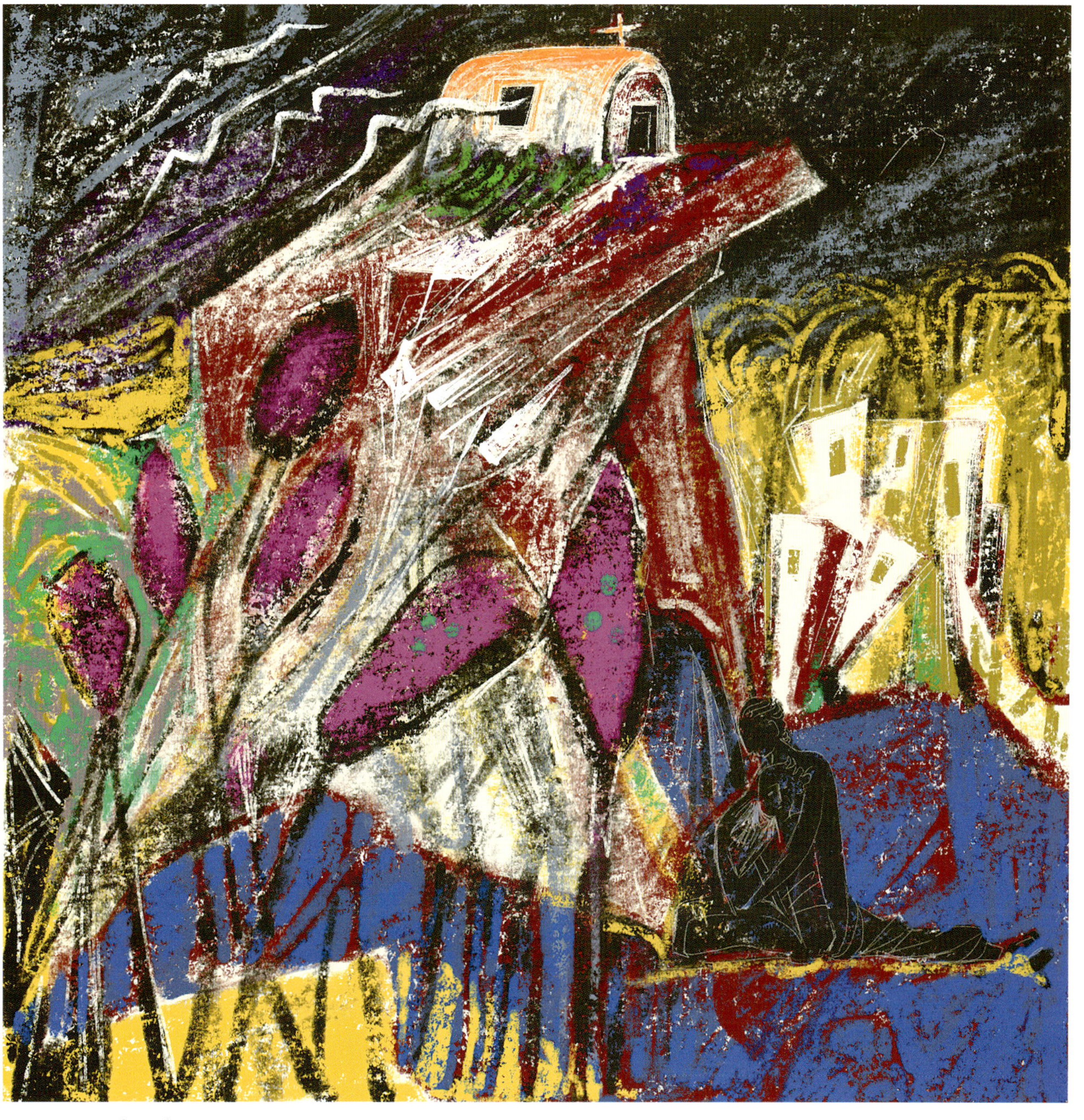

A CHURCH I, digital painting, 140 x 140 cm

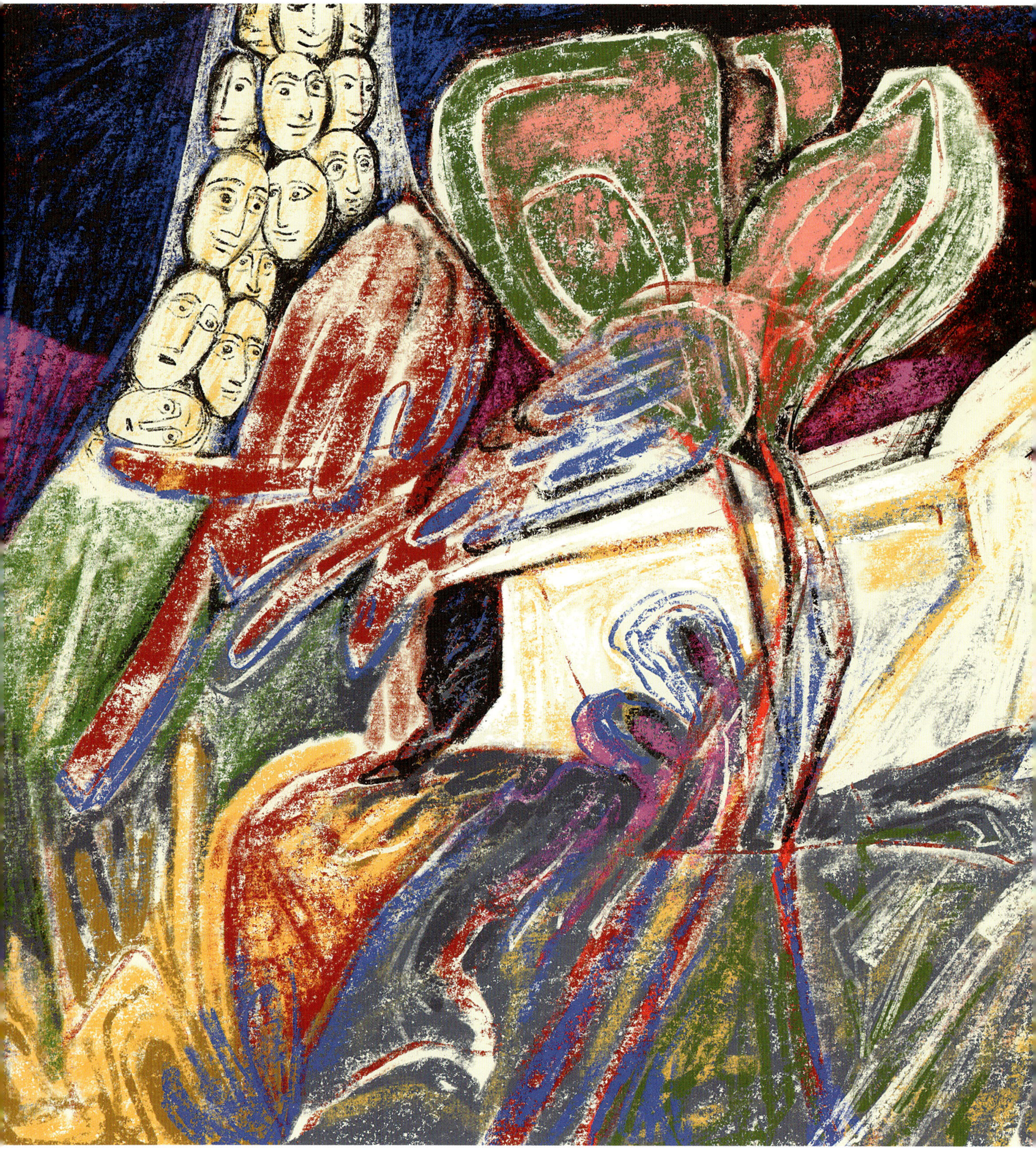

A CHURCH III, digital painting, 140 x 280 cm

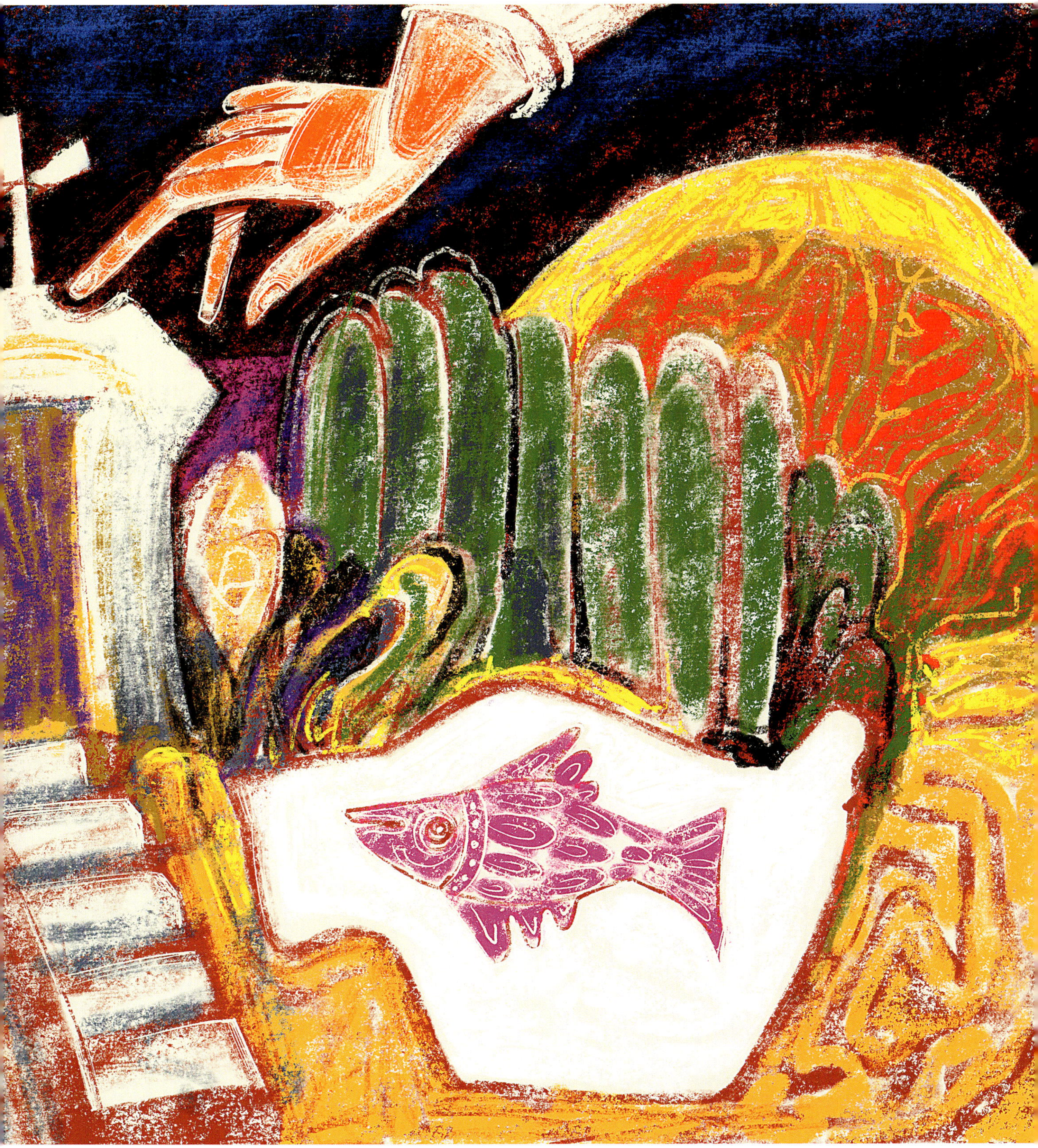

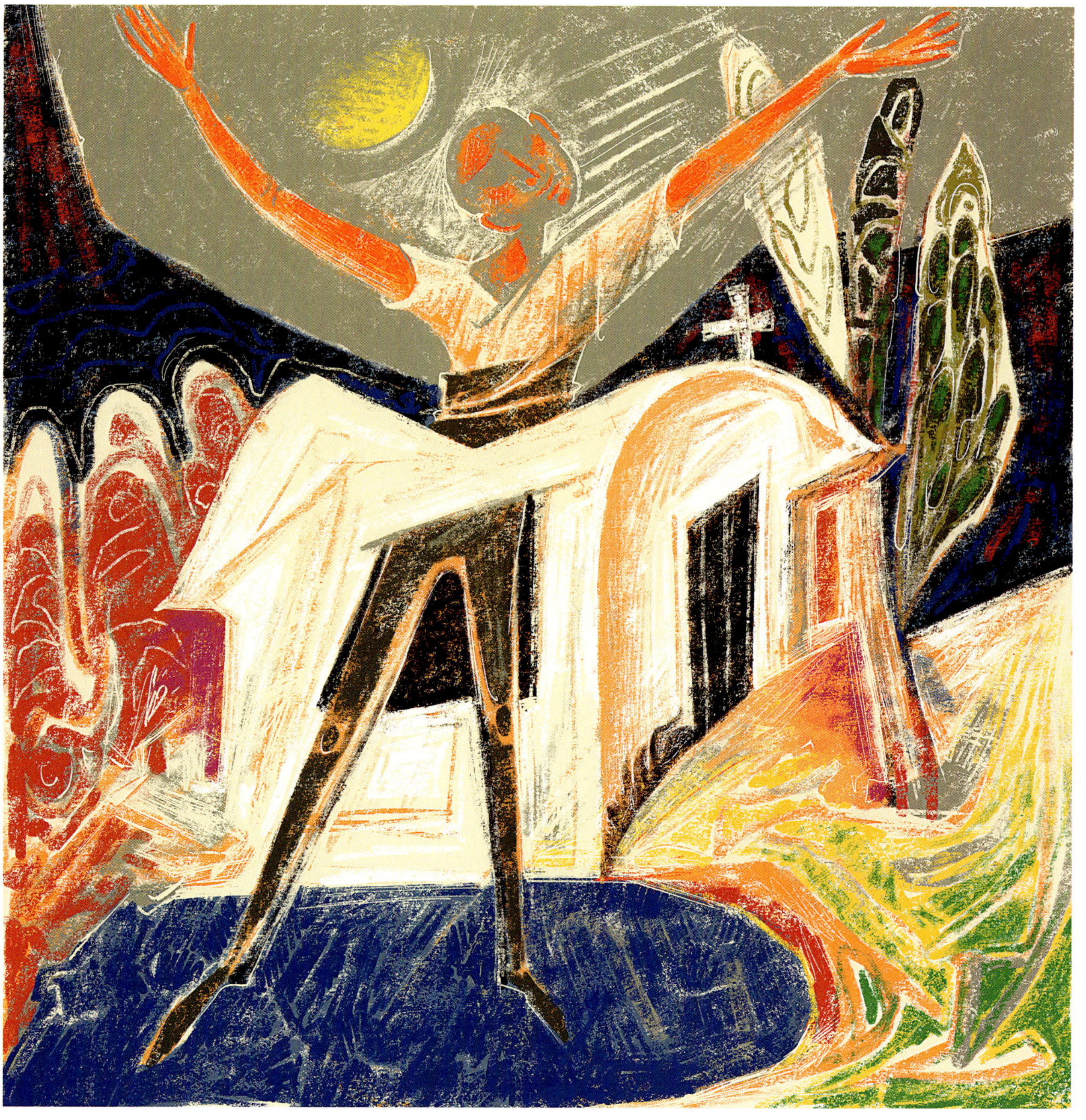

A CHURCH II, digital painting, 140 x 140 cm

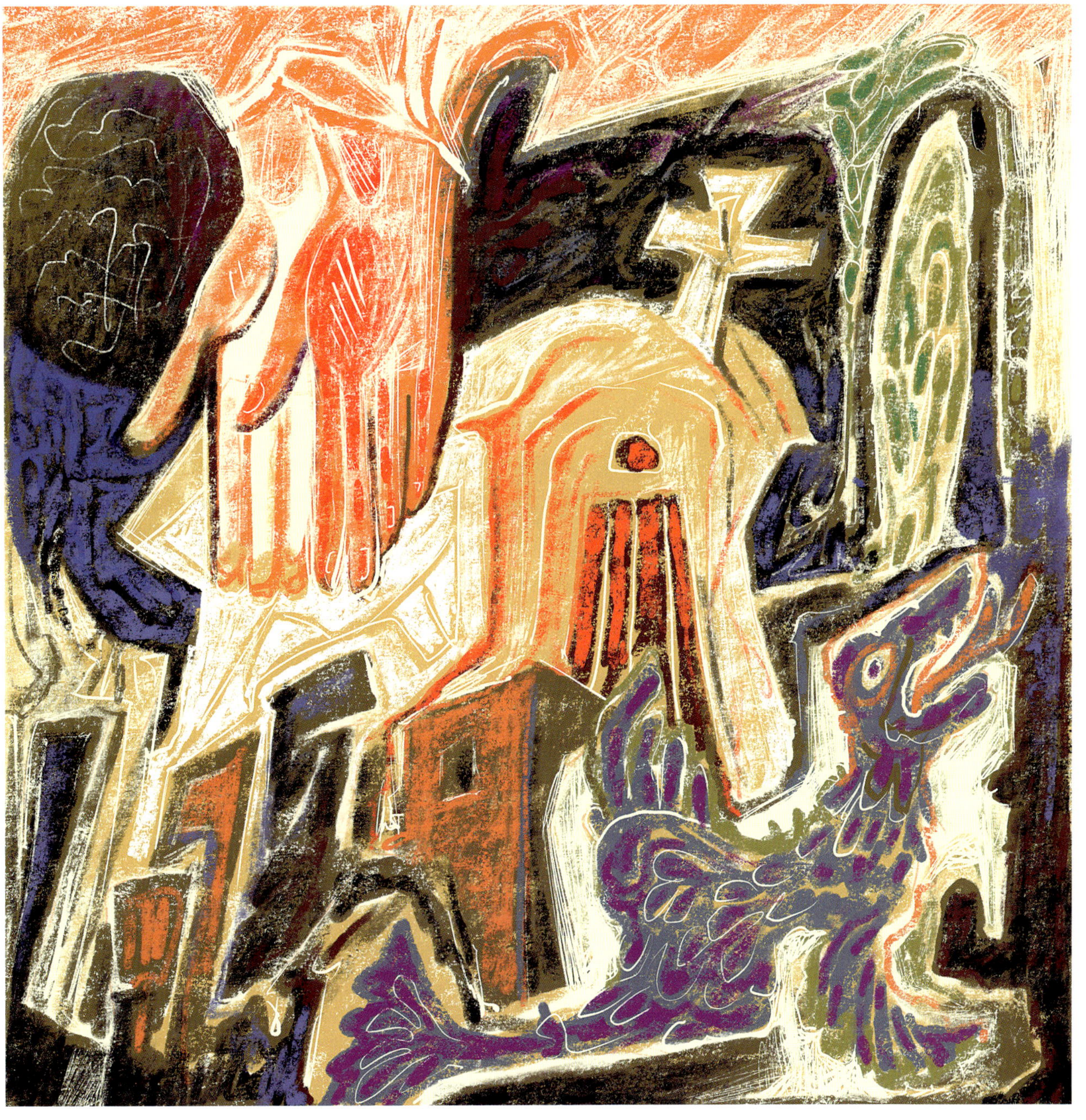

A CHURCH IV, digital painting, 140 x 140 cm

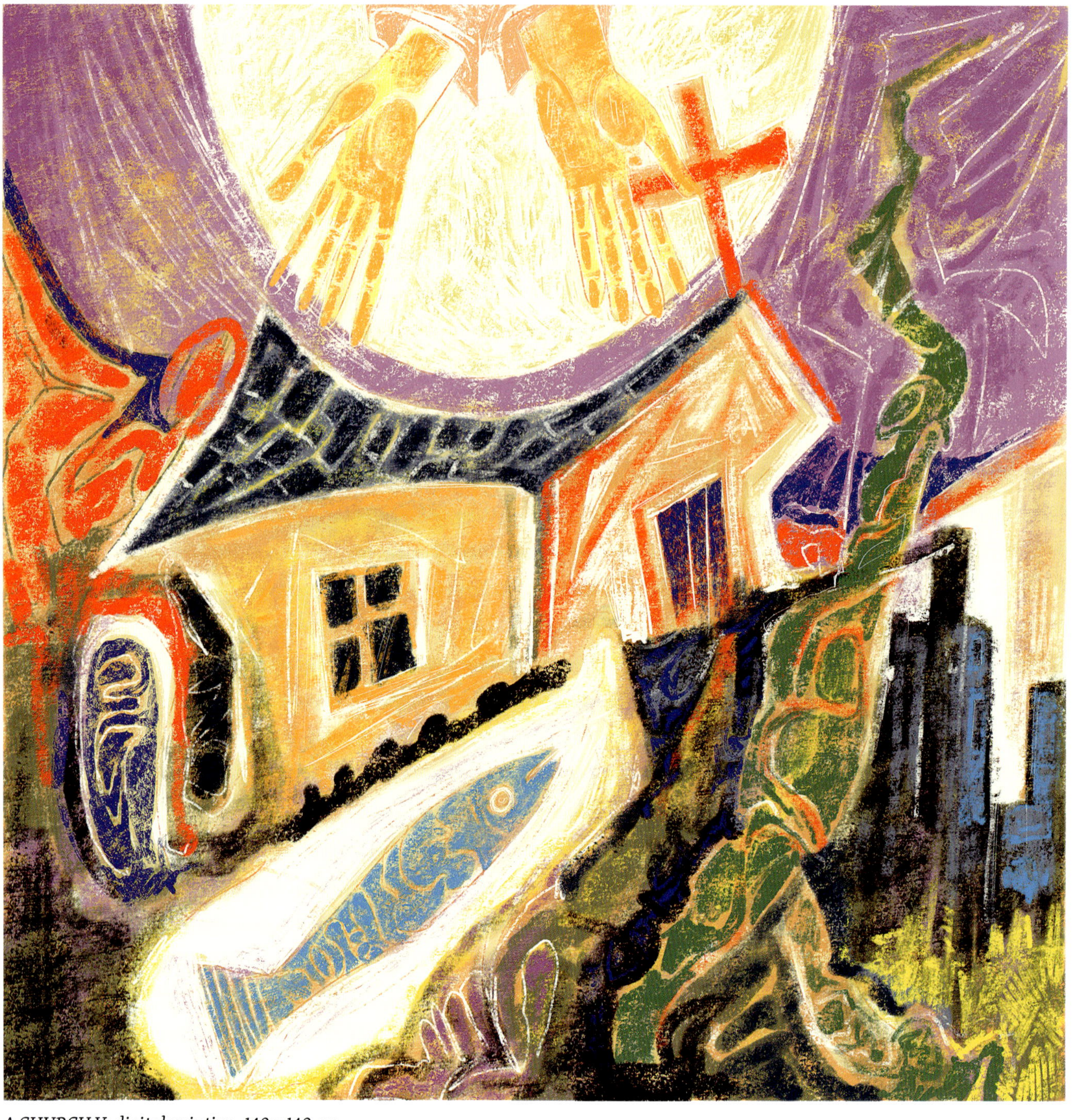

A CHURCH V, digital painting, 140 x 140 cm

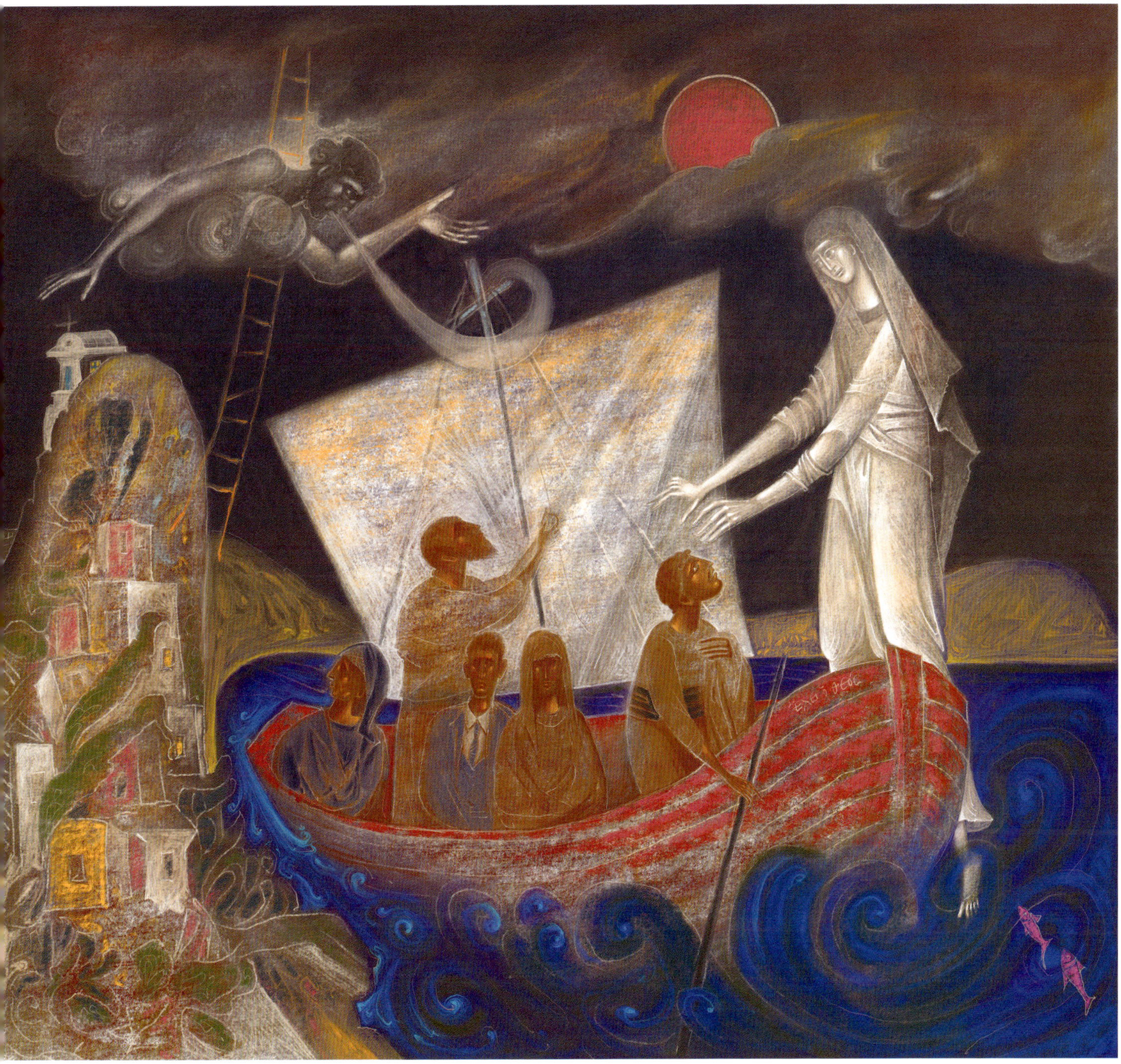

EROS, HERO, PAPADIAMANTIS,
digital painting, 140 x 140 cm

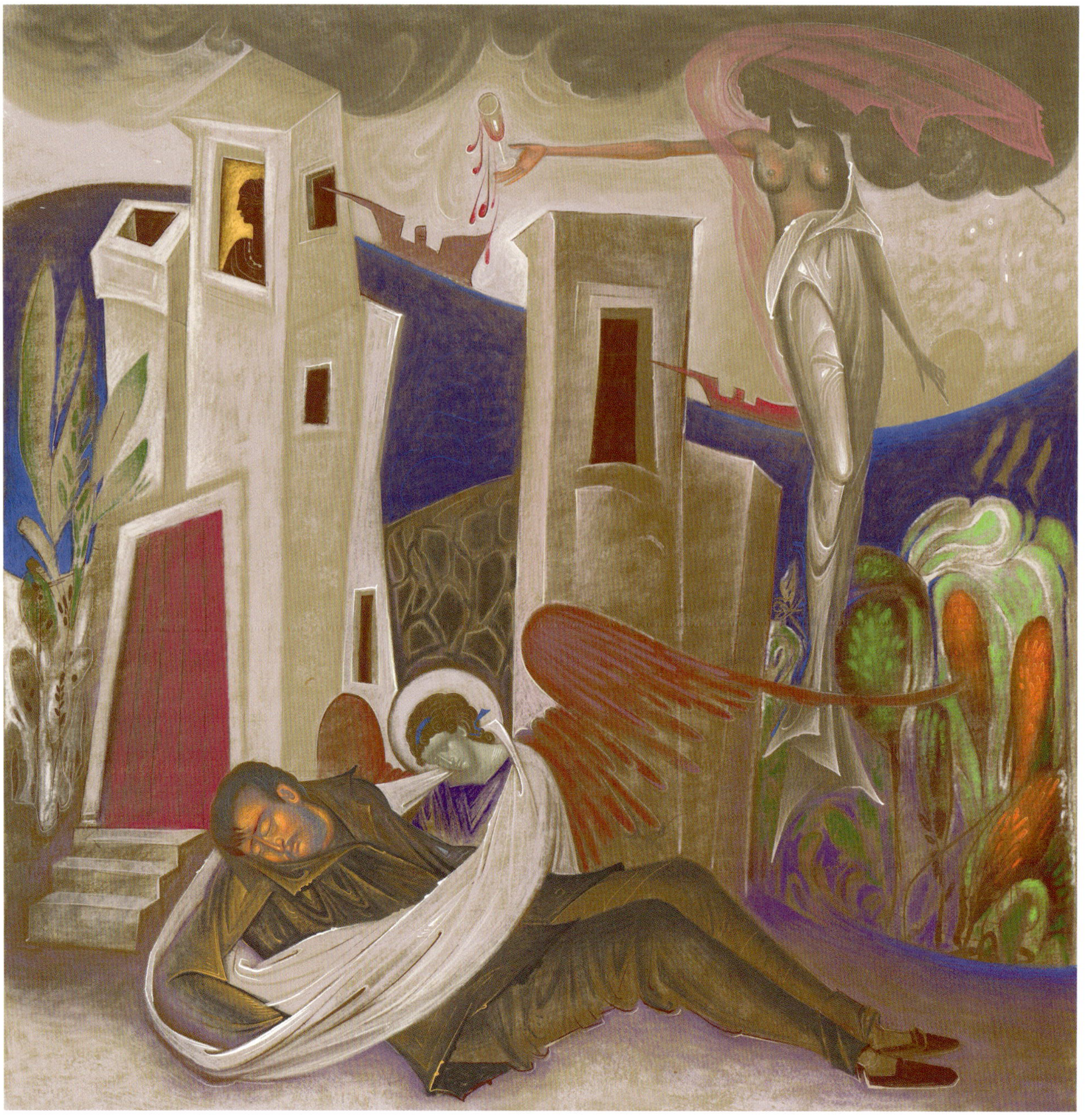

LOVE IN SNOW, PAPADIAMANTIS,
digital painting, 140 x 140 cm

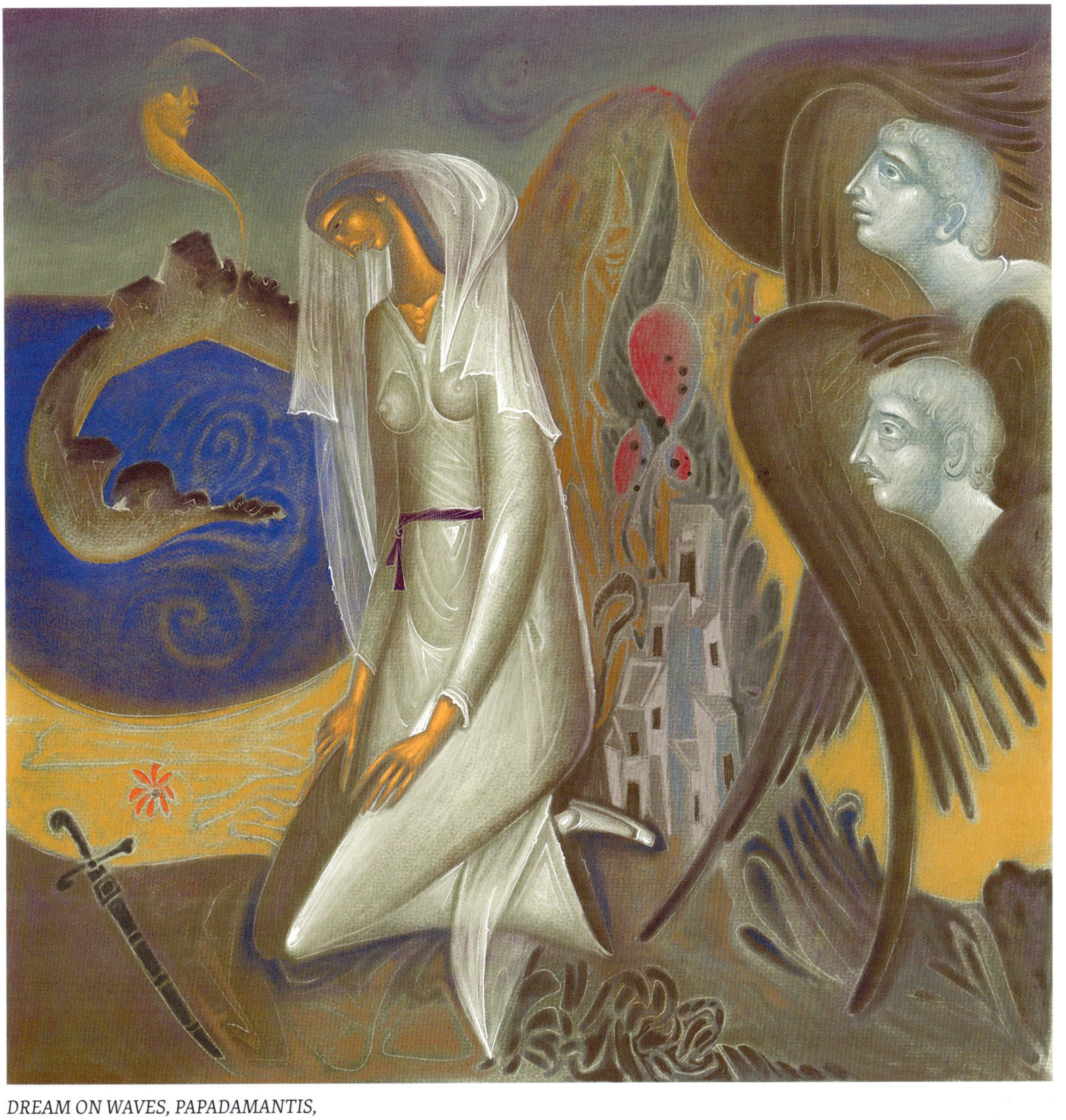

DREAM ON WAVES, PAPADAMANTIS,
digital painting, 140 x 140 cm

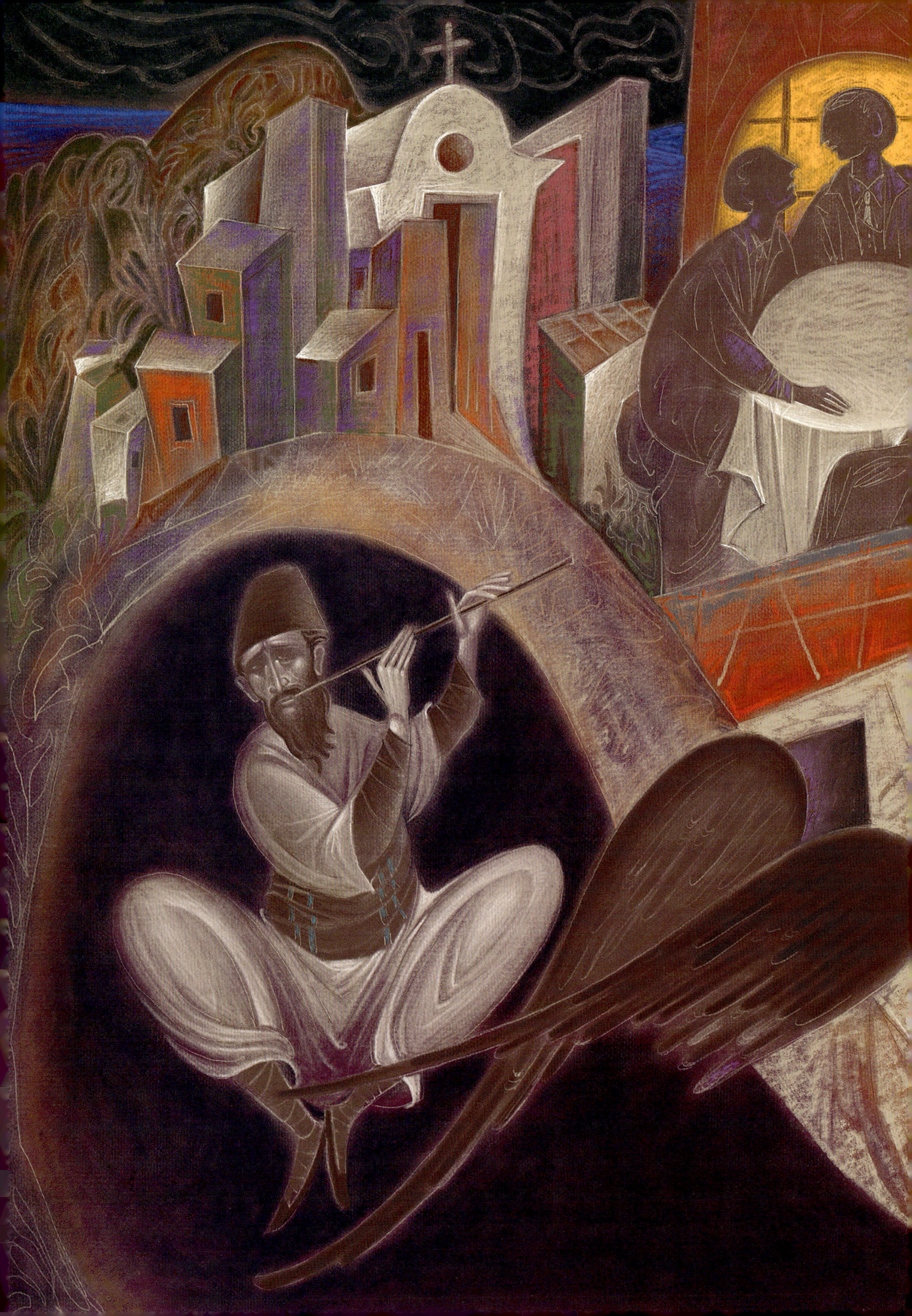

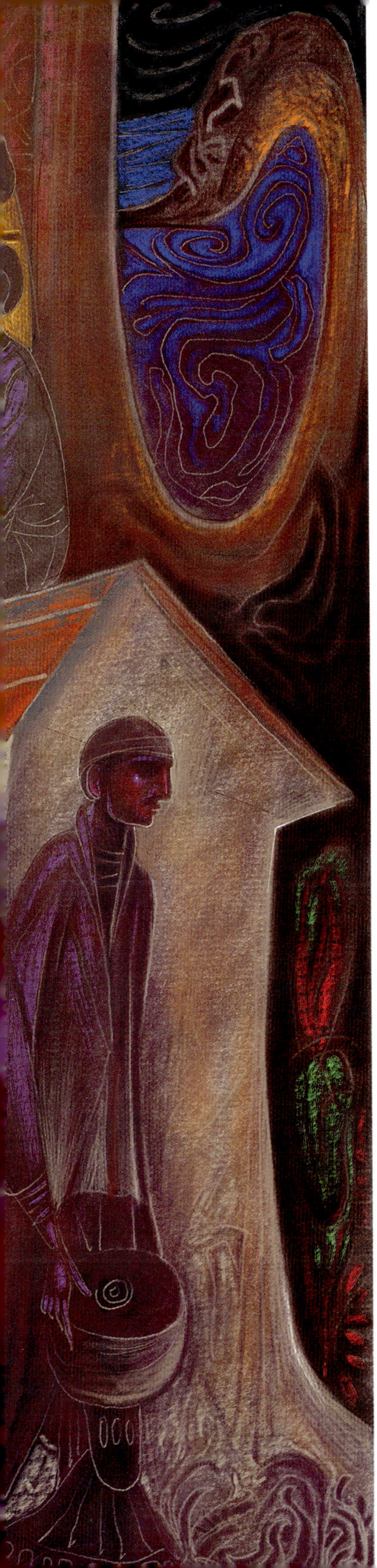

THE FALLEN DERVISH, PAPADIAMANTIS,
digital painting, 140 x 140 cm

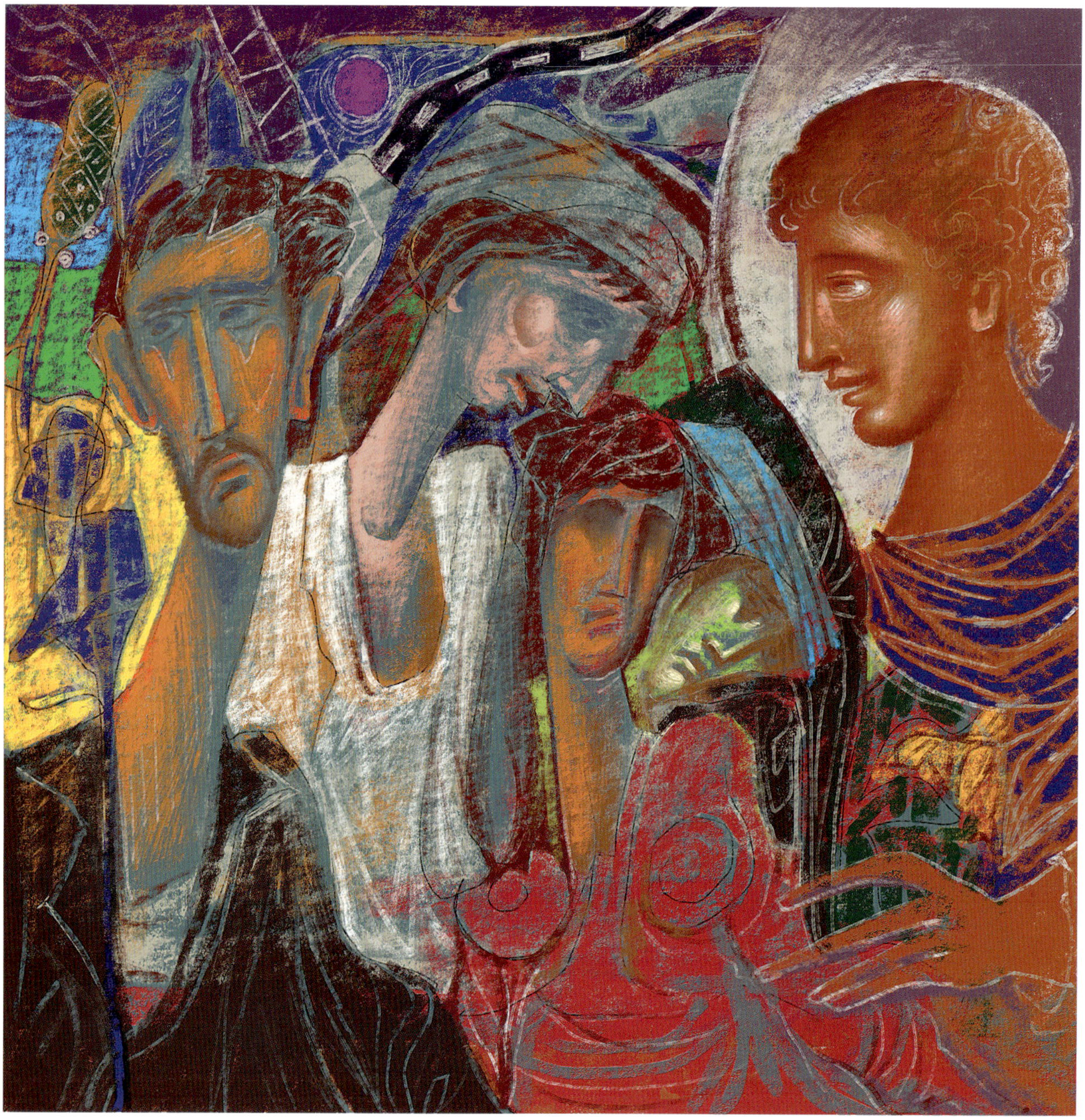

TARKOVSKY. MIRROR,
digital painting, 120 x 120 cm

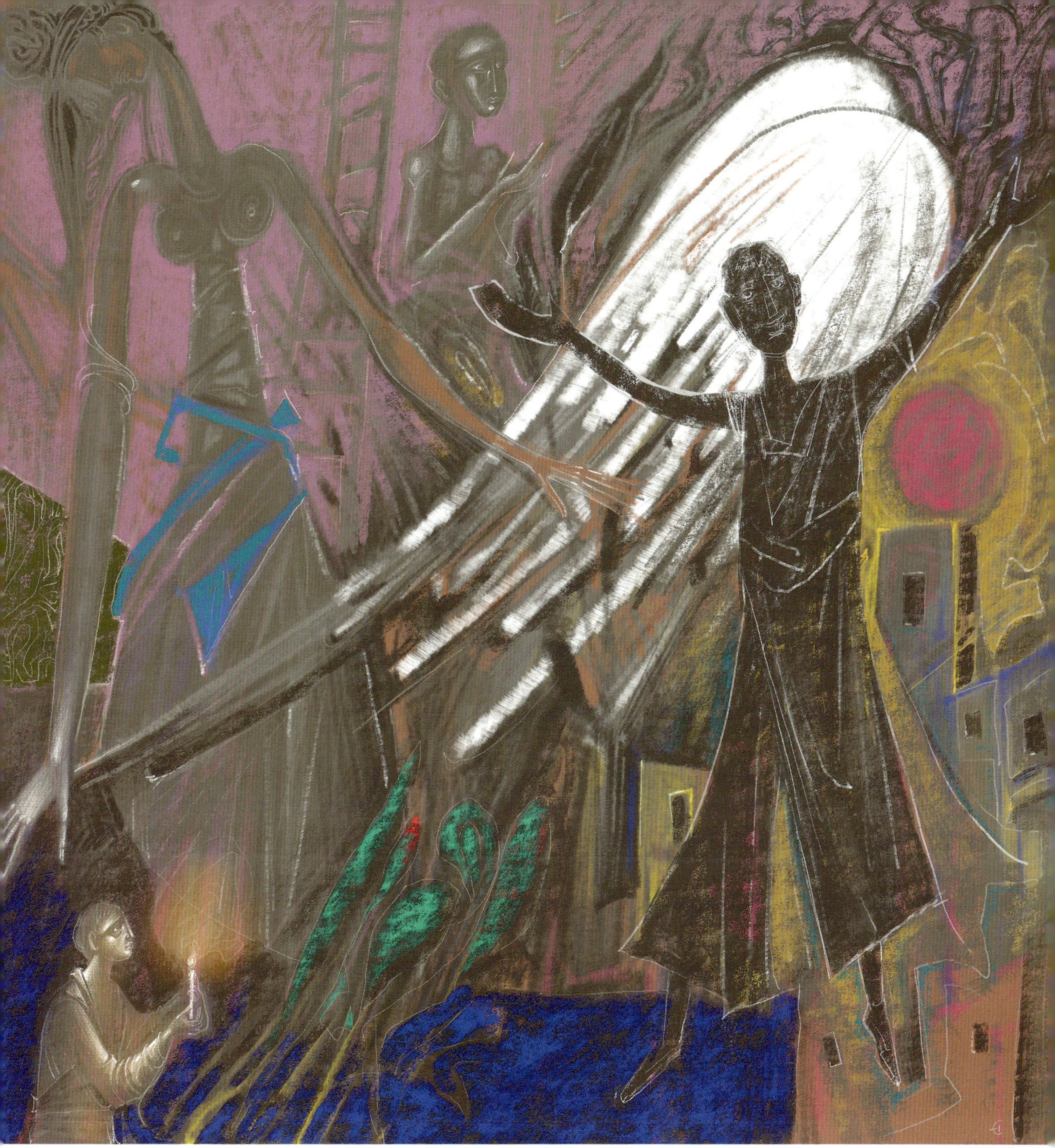

TARKOVSKY. NOSTALGIA,
digital painting, 120 x 120 cm

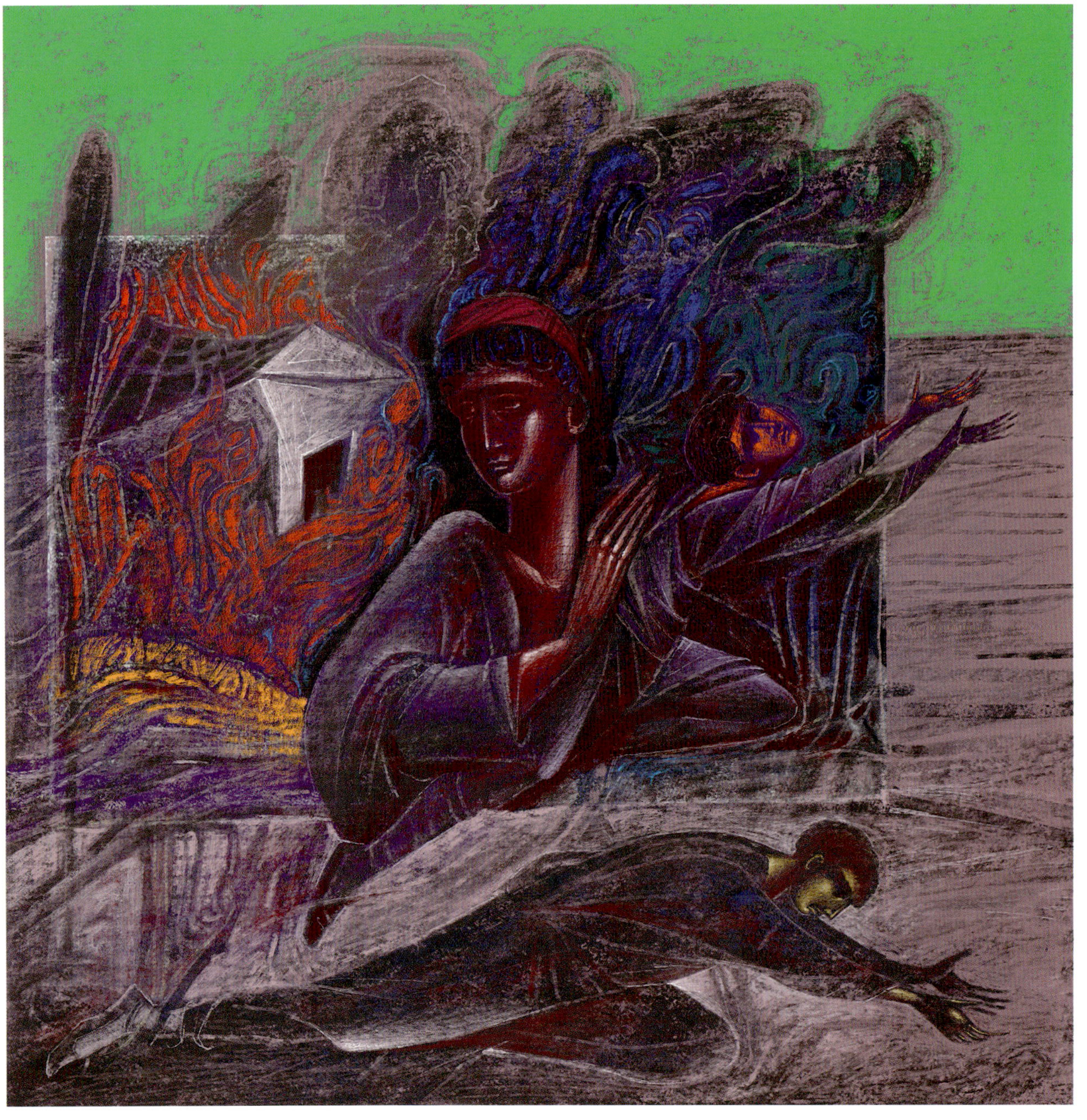

TARKOVSKY. SACRIFICE,
digital painting, 120 x 120 cm

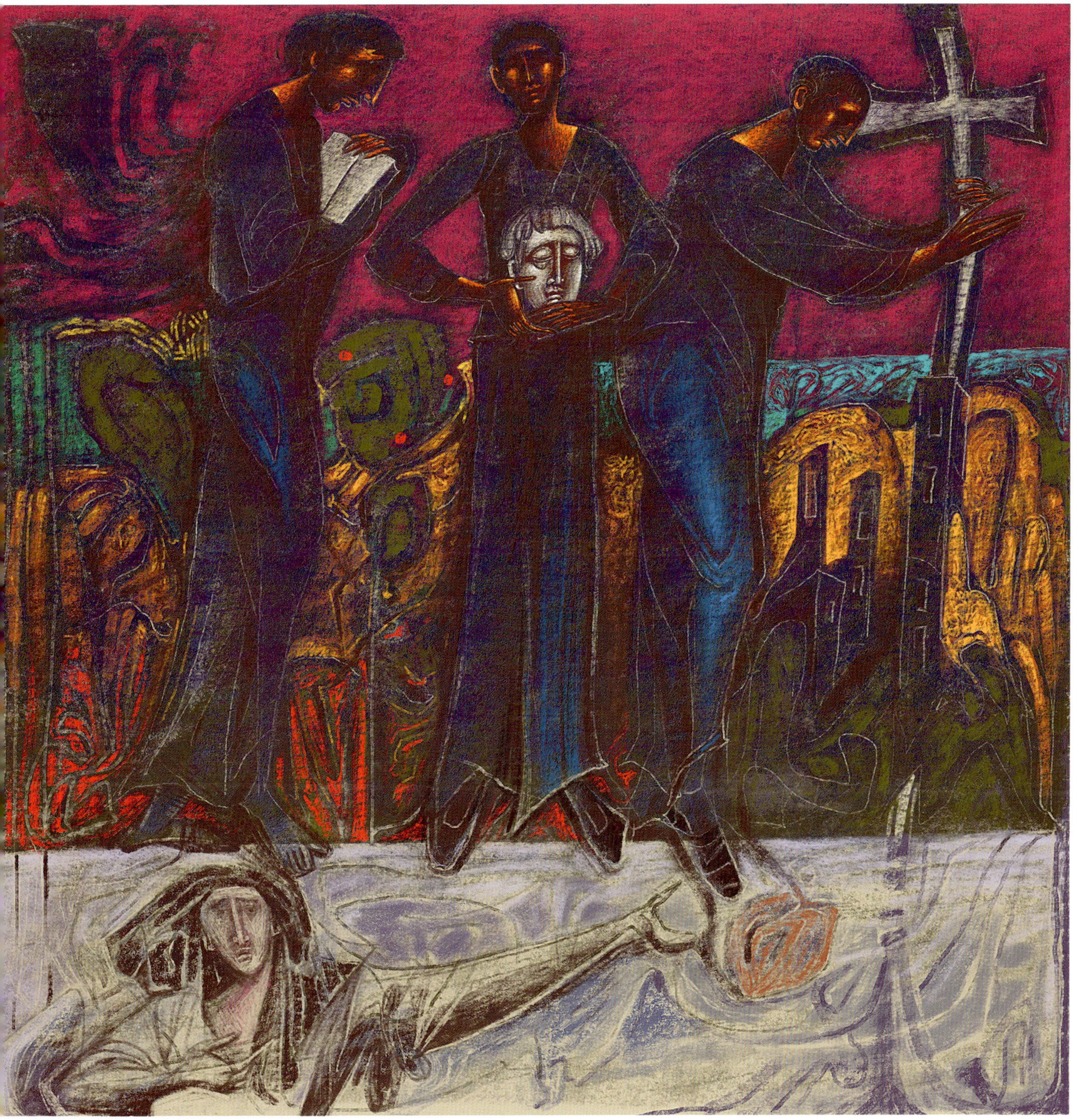

TARKOVSKY, STALKER,
digital painting, 120 x 120 cm

Notes on Authors

Annemarie Weyl Carr earned her BA from Swarthmore College and her MA and PhD from the University of Michigan. She is Professor Emerita of Art History at Southern Methodist University, where she held the title of University Distinguished Professor. A specialist in Byzantine art, her scholarship has focused on icons, cultural exchange in the Crusader-era Levant—especially Cyprus—and medieval women artists. She is the author of several books, including *Byzantine Illumination, 1150–1250, A Masterpiece of Byzantine Art Recovered, and Cyprus and the Devotional Arts of Byzantium*. A former editor of *Gesta* and past president of the International Center of Medieval Art, she has also served as a trustee of the Cyprus American Archaeological Research Institute. Her many teaching awards include the College Art Association's Lifetime Achievement Award (2006).

Peter C. Bouteneff is Professor of Systematic Theology at St. Vladimir's Orthodox Theological Seminary, where he also directs the Institute of Sacred Arts. A theologian and musician, he explores the intersections of theology, art, and culture. He studied at Yale and Oxford and worked closely with Metropolitan Kallistos Ware. Among his publications are *Sweeter than Honey: Orthodox Thinking on Dogma and Truth* and *How to Be a Sinner*. His work regularly engages the visual arts, and he has curated exhibitions exploring the role of sacred art in contemporary contexts.

Judith Wolfe, born in Vienna, studied in Jerusalem and Oxford, and taught in Berlin and Oxford before joining the University of St Andrews in 2014. A scholar in systematic and philosophical theology, she focuses on eschatology and its broader implications in theology, anthropology, and epistemology. She recently published The Theological Imagination (Cambridge University Press) and leads major research projects including Widening Horizons in Philosophical Theology and Art as Revelation. She is editor of the *Cambridge Companion to Eschatology*, co-editor of the Oxford History of Modern German Theology, and founding editor of the *Journal of Inklings Studies*. Her previous books include *Heidegger's Eschatology, Heidegger and Theology*, and *C.S. Lewis and His Circle*.

Justin Willson is an art historian specializing in Byzantine art and visual culture. His research explores the intersection of aesthetics, theology, and the body, particularly in late Byzantium. He earned his Ph.D. from Princeton University, where his dissertation examined the role of abstraction and spirituality in Palaiologan art. His work also delves into color theory, sacred space, and the performative aspects of iconography. In addition to academic writing, he contributes to curatorial projects and teaches courses on Orthodox art, liturgy, and the history of Christian aesthetics.

Niobe Kritikou Born in Athens in 1993, she graduated with distinction from the Athens School of Fine Arts, completing her studies in History and Theory of Art and Visual Arts. Her career spans between creating and curating exhibitions, with participations and collaborations in festivals, museums and institutions such as Documenta 14 and EMST. As Artistic Director of the Arts Centre Mets, she organizes exhibitions in Greece and internationally, collaborating with both emerging and established artists. Her aim is to connect contemporary art with the public, highlighting the relationship of Greece's cultural heritage with new creative approaches and international dialogues.

Bishop Maxim Vasiljević of Los Angeles and Western America earned his doctorate in dogmatics and patristics from the University of Athens in 1999. He completed post-doctoral work in Byzantine History and Theology at the Sorbonne in Paris and enrolled at painting classes at the French Academy of Fine Arts. He teaches at Holy Cross Greek Orthodox School of Theology in Boston. He is also an accomplished painter with exhibitions and iconography classes worldwide. His notable works include *History, Truth, Holiness* (2011), *Theology as a Surprise* (2018), *Wonder as the Beginning of Faith* (2022), and *Saved by Beauty: Dostoevsky and America* (2022). He is one of the founder of Metropolitan John Zizioulas Foundation.

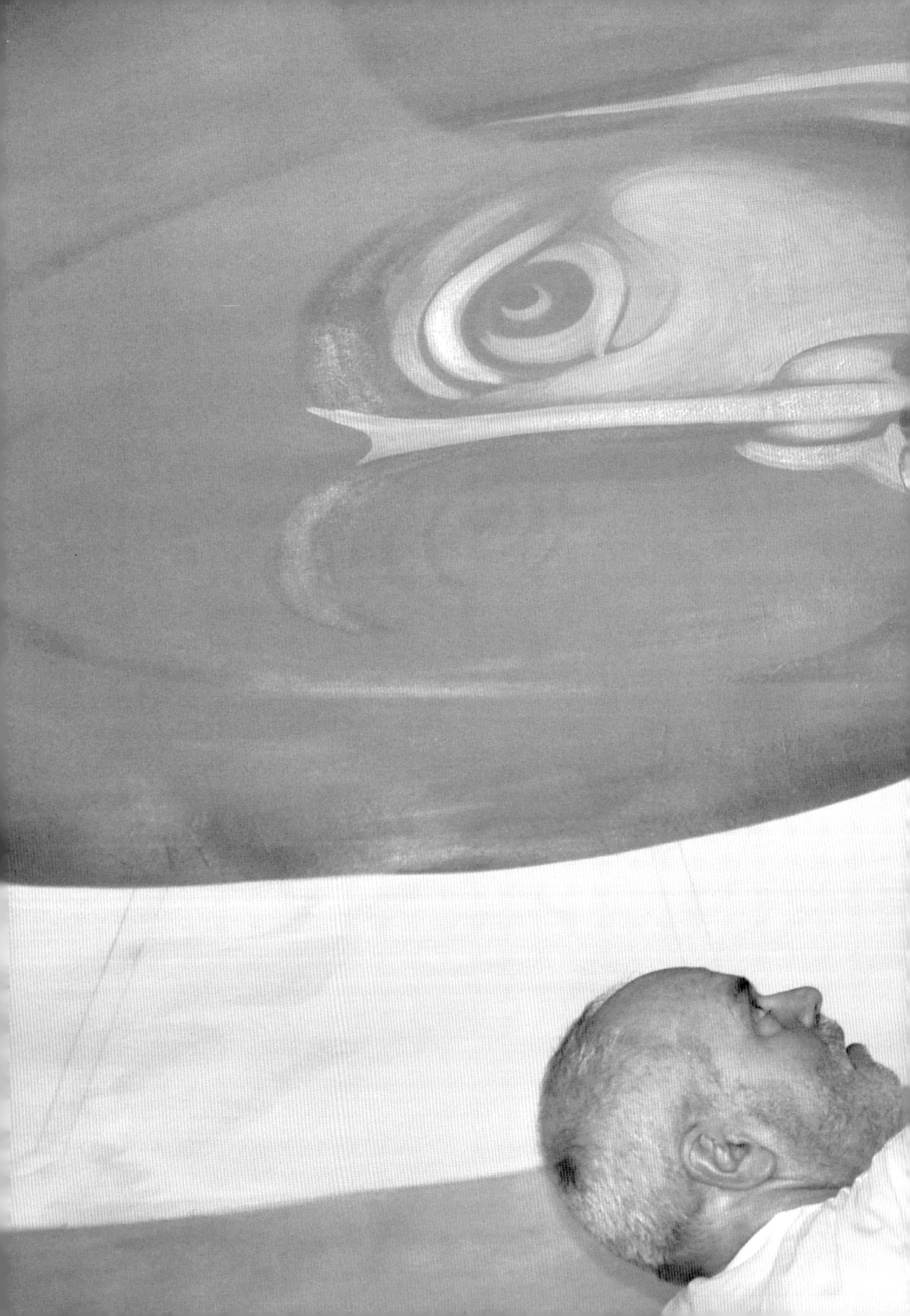

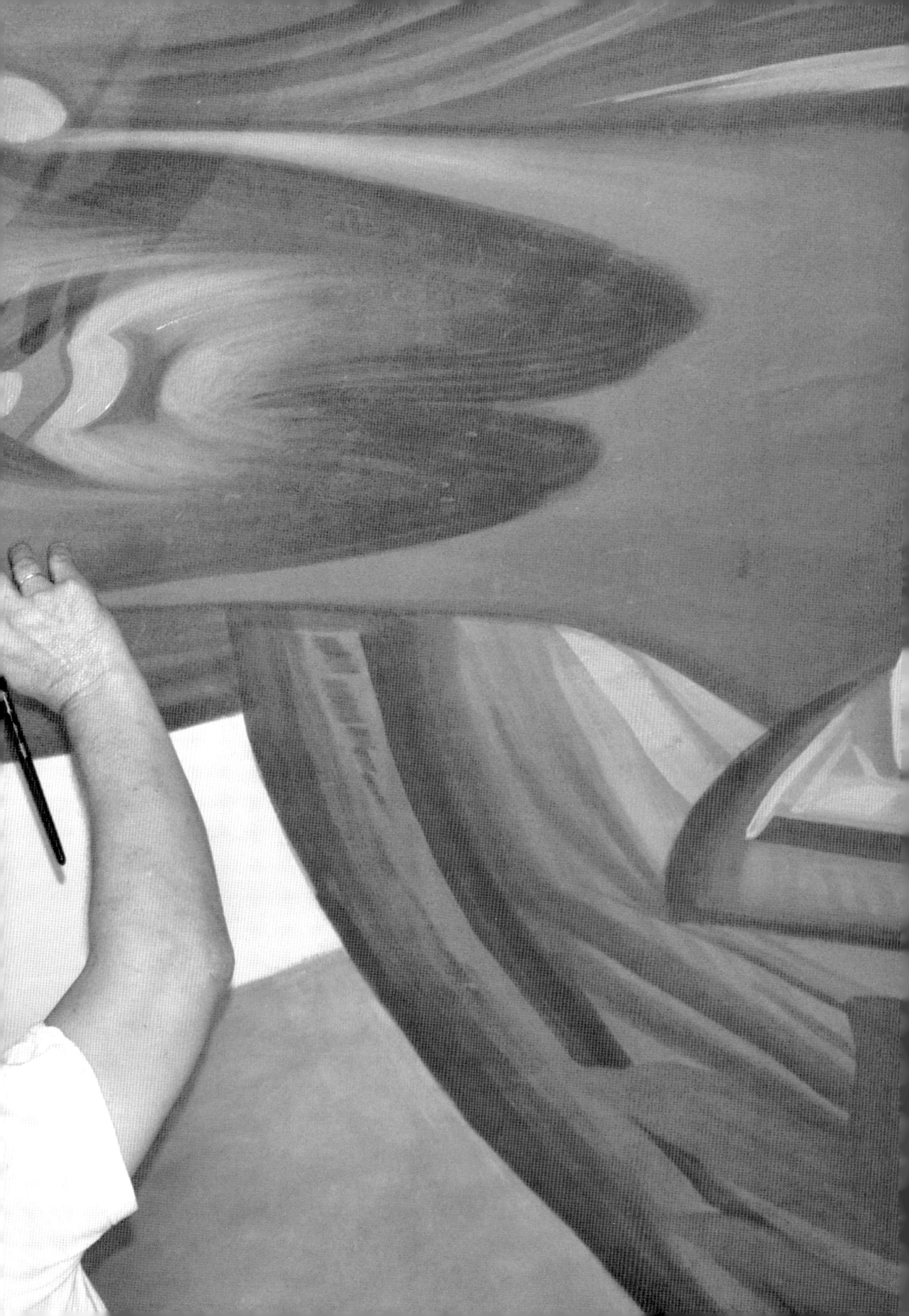

Here ends the book In the Joy of Cinnabar, which gathers the luminous work and vision of George Kordis, edited with devotion by Bishop Maxim Vasiljević. Design and Prepress by Denis Vikić. Published by St. Sebastian Orthodox Press, as part of the series Contemporary Christian Thought, number 104, and printed with craftsmanship by Birograf COMP, in Belgrade, Serbia. Completed on the first day of April, in the year of our Lord 2025.